PRESENCE

The Foundation of

EMOTIONAL INTELLIGENCE

By

Joseph Anand

Foreword

Presence is not a technique.
It is a way of being.

Long before we learned to optimize our calendars, quantify our performance, or accelerate our decisions, presence was the quiet human capacity that made leadership trustworthy, relationships meaningful, and life intelligible. We sensed it instinctively - when someone truly listened, when a room settled because a calm mind entered it, when a difficult moment was met without reaction but with awareness.

And yet, presence has slowly disappeared from modern life.

We live in an age of unprecedented speed. Information travels faster than reflection. Reaction often precedes understanding. Productivity is rewarded more than clarity, and busyness is mistaken for importance. In this environment, emotional intelligence is not lost - but it is crowded out. The cost is subtle at first: strained relationships, decision fatigue, shallow attention. Over time, it becomes profound: leadership without grounding, communication without connection, achievement without fulfillment.

This book is an invitation to return.

Presence is not about slowing down for its own sake, nor about withdrawing from ambition or responsibility. It is about reclaiming the inner stability that allows us to engage fully - with people, pressure, and purpose - without being consumed by them. Presence is the foundation upon which emotional intelligence stands. Without it, self-awareness becomes delayed, empathy becomes reactive, and regulation becomes effortful. With it, clarity emerges naturally.

The ideas in this book were shaped not only by research and professional experience, but by life itself—by moments of challenge, loss, leadership, and learning where presence was the difference between reaction and wisdom. You will not find quick fixes here. You will find practices that are simple, but not simplistic. Reflections that are practical, but deeply human. This book does not ask you to become someone new. It asks you to return to what has always been available - your capacity to notice, to choose, and to respond with intention.

Whether you are a leader navigating complexity, a professional under constant pressure, or a human being seeking steadiness in a noisy world, this journey begins in the same place: with attention.

Presence is the gap where emotional intelligence lives.
And in that gap, everything changes.

Dedication

This book is dedicated to the quiet moments that changed my life.

To my late wife, Bibiana -
You showed me that emotional intelligence is not something we practice, but something we *embody*. Long before I had words for presence, you lived it. In your way of listening, people felt received. In your calm, urgency softened. In your compassion, dignity was preserved - especially when circumstances were difficult.
You taught me that wisdom does not hurry, that strength does not need volume, and that love often reveals itself in silence. Though you are no longer here in form, your presence continues to guide me. This book carries your imprint in every space between the words.

To my children, Desmond and Cynthia -
You taught me resilience, humility, and the courage to remain open when life invites closure. Through you, I learned that presence is not withdrawal from life, but deeper participation in it. Your patience with my learning, your quiet strength through loss, and your trust in my journey reminded me that the greatest gift we offer one another is our full attention. This book exists because of you - and is for you.

To the teachers, mentors, coaches, and companions who modeled calm in complexity—
You showed me that leadership begins within, that clarity emerges when we slow down, and that the most powerful

interventions are often invisible. You helped me understand
that presence is not passivity, but wisdom in motion.

To those living in a world that never stops -
This book is for you.
For the professionals carrying relentless pressure.
For the leaders navigating uncertainty.
For the parents, caregivers, creators, and seekers who feel the
pull of speed yet sense its cost.
You are not behind. You are not failing. You are human in a
system that rarely pauses long enough to notice humanity.

This book is an invitation -
to stop without quitting,
to slow without falling behind,
and to return to the place where emotional intelligence truly
begins - Presence.

Table of Contents:

Book Overview

What Is *Presence* About?

Presence: The Foundation of Emotional Intelligence is about reclaiming the most overlooked human capacity in modern life: the ability to be fully here.

While emotional intelligence is widely discussed, taught, and measured, this book makes a clear and essential argument—**emotional intelligence cannot function without presence**. Presence is the ground from which self-awareness, regulation, empathy, and wise action emerge. Without it, emotional intelligence becomes mechanical, performative, or fragile under pressure.

This book explores why so many capable, intelligent, and well-intentioned people feel emotionally exhausted, reactive, distracted, or disconnected—not because they lack skills, but because they live in a near-constant state of absence.

The Central Premise

We live in an age of speed, distraction, and emotional overload. Attention is fragmented. Nervous systems are over-stimulated. Reactions replace responses.

In this environment:

- Emotions are felt but not processed
- Awareness is conceptual rather than embodied
- Regulation turns into suppression
- Empathy becomes effortful and draining

Presence reframes emotional intelligence not as something we **do**, but as something that naturally arises when we are **present**.

What the Book Does

This book:

- Names **presence** as the missing foundation beneath emotional intelligence
- Explains how modern life trains absence - and the emotional cost of it
- Connects presence to the nervous system, attention, and emotional regulation
- Shows how reactivity, burnout, and disconnection are symptoms of lost presence, not personal failure
- Guides readers toward restoring presence as a lived state, not a technique

Rather than offering quick fixes or motivational tactics, the book invites a deeper shift -from managing emotions to **meeting experience directly**.

What This Book Is *Not*

This is not:

- A productivity book
- A mindfulness manual filled with techniques
- A positivity or motivation guide
- A call to slow life down or withdraw from the modern world

Presence, as presented here, is not about doing less.
It is about **being fully available to what you are already doing**.

Who This Book Is For

This book is written for:

- Leaders who feel emotionally stretched despite competence
- Professionals navigating pressure, complexity, and constant demands
- Educators, coaches, and facilitators of emotional intelligence
- Anyone who senses that something essential is missing beneath performance and success

It speaks especially to those who are functioning well externally yet feel fragmented internally.

The Promise of the Book

Presence offers a quiet but powerful promise:

When presence is restored:

- Awareness deepens without effort
- Regulation becomes natural
- Empathy stabilizes
- Clarity returns under pressure
- Emotional intelligence becomes sustainable

This book is not about becoming better.
It is about becoming **here**.

From that place, emotional intelligence stops being a skill to practice—and starts becoming a way of being.

Introduction: The Missing Foundation Beneath Emotional Intelligence

Emotional intelligence is everywhere.

It appears in leadership frameworks, performance reviews, hiring criteria, education systems, and self-development programs. We are encouraged to become more self-aware, regulate our emotions, communicate empathetically, and lead with emotional intelligence. The language is familiar. The intention is sound.

And yet, something essential is missing.

Despite decades of research and widespread adoption, emotional intelligence often fails at the very moments it is most needed—under pressure, in conflict, during uncertainty, and when emotions run high. People who understand emotional intelligence conceptually still react impulsively. Leaders trained in empathy still become defensive. Highly capable professionals still lose clarity when stakes rise.

The problem is not a lack of knowledge.
It is a missing foundation.

Most approaches to emotional intelligence focus on *what* to do:

- Notice your emotions

- Manage your reactions
- Communicate effectively
- Show empathy
- Make thoughtful choices

But they rarely address the more fundamental question:

From what inner state are you doing all of this?

Emotional intelligence does not begin with skills or techniques. It begins with the capacity to be present - present with oneself, with others, and with what is happening in the moment. Without presence, emotional intelligence becomes effortful, inconsistent, and fragile. With presence, it becomes natural.

Presence is the ground beneath emotional intelligence.
And yet, it is rarely named, taught, or developed directly.

When presence is absent, self-awareness becomes distorted. We think we know what we feel, but we are already reacting. Regulation turns into suppression or control. Empathy becomes draining because there is no internal boundary. Communication becomes performative rather than authentic. Decisions are made from urgency rather than clarity.

Presence is not a personality trait.
It is not calmness.
It is not meditation or slowness.
It is not spiritual language.

Presence is the ability to remain internally grounded and aware in the midst of emotion, pressure, and complexity. It is the inner stability that allows emotions to be felt without overwhelming us, thoughts to be observed without being believed automatically, and situations to be engaged without losing ourselves.

This book makes a simple but powerful claim:

Emotional intelligence is not something you apply.
It is something that emerges - when presence is established.

Rather than adding more tools, this book goes beneath technique to state. Rather than asking how to manage emotions, it asks how to meet them without collapse or control. Rather than teaching performance, it cultivates inner conditions.

Presence is not about becoming better, calmer, or more emotionally skilled. It is about becoming more available— internally steady enough to experience life directly, respond wisely, and lead without reactivity.

The chapters that follow will explore why presence has eroded in modern life, how it functions within the nervous system, and how it gives rise to self-awareness, regulation, empathy, and clear decision-making. You will not be asked to withdraw from life or adopt rigid practices. Presence is not something added on—it is something remembered.

Emotional intelligence has long been treated as the destination.
This book invites you to discover its true beginning.

Presence comes first.

Chapter 1 — The Cost of Living Unpresent

Distraction, Speed, and Emotional Fragmentation

We rarely notice when presence leaves us.
It does not announce its departure.
It fades quietly—replaced by urgency, interruption, and mental noise.

Most people do not wake up intending to live unpresent. They wake up intending to be productive, responsive, efficient. Yet by the end of the day, many feel oddly absent from their own lives—tired without having done anything meaningful, connected yet lonely, informed yet unclear.

This chapter explores that invisible cost.

The Age of Constant Elsewhere

Modern life trains attention to live everywhere except where the body is.

We scroll while eating.
We plan while listening.
We respond while feeling.

Attention is constantly pulled forward into what's next or sideways into what's elsewhere. Rarely does it rest fully in what *is*.

The result is not simply distraction—it is *dislocation*.
The mind leaves the moment. The body remains. Emotions are left unmanaged in between.

Presence is not lost because we are careless.
It is lost because speed has become the default operating system.

Speed as a State, Not a Strategy

Speed was once a tool. Now it is a state.

We move quickly even when nothing requires urgency.
We think rapidly even when reflection is needed.
We speak fast, decide fast, react fast—often without noticing what we feel.

This matters because emotions do not operate at the speed of machines.

They require sensing.
They require space.
They require attention.

When life moves faster than emotional processing, emotions do not disappear. They fragment.

Emotional Fragmentation: The Hidden Tax

Emotional fragmentation occurs when feelings are experienced without being integrated.

A frustration felt but not acknowledged.
An anxiety managed through distraction.
A sadness postponed until later—then forgotten, but not resolved.

Over time, this creates a scattered inner landscape. People feel "off" without knowing why. They overreact to small triggers. They struggle to articulate what they feel. They mistake emotional numbness for calm.

This is not emotional weakness.
It is emotional overload without presence.

Distraction Is Not Neutral

Distraction is often framed as harmless or even relaxing. But distraction is not neutral—it diverts attention away from emotional signals that need awareness.

When attention is constantly elsewhere:

- Emotions are delayed rather than processed
- Stress accumulates beneath consciousness
- Empathy becomes effortful
- Self-awareness weakens

Distraction may feel like relief, but it often functions as avoidance. What we do not attend to does not disappear—it waits.

The Cost We Normalize

Living unpresent comes with costs we have learned to accept as normal:

- Chronic tiredness without clear cause
- Irritability that feels disproportionate
- Difficulty listening deeply
- Shallow satisfaction even after success
- A persistent sense of being "behind"

These are not signs of personal failure. They are symptoms of sustained absence.

Presence is not lost in dramatic moments—it is eroded in ordinary ones.

When Emotional Intelligence Has No Ground to Stand On

Emotional intelligence is often taught as a set of skills: self-awareness, regulation, empathy, communication. But skills require a foundation.

That foundation is presence.

Without presence:

- Awareness becomes conceptual, not lived
- Regulation becomes suppression
- Empathy becomes performance
- Communication becomes reactive

Presence is the ground from which emotional intelligence functions. When the ground is unstable, the structure cannot hold.

The Quiet Longing Beneath the Noise

Beneath distraction and speed lies a quiet longing many people share but rarely name—the desire to feel *here*.

To be with a conversation without drifting.
To experience emotion without being overwhelmed.
To act from clarity rather than compulsion.

This longing is not for a slower life, but for a more *inhabited* one.

Reclaiming the First Capacity

Presence is not something to add to life.
It is something to restore.

Before strategies.
Before techniques.
Before emotional mastery.

The cost of living unpresent is not just stress or fatigue—it is the gradual separation from oneself.

And that separation is reversible.

The next chapter explores how absence became normalized— and why reclaiming presence is not a luxury, but a necessity for emotional intelligence in modern life.

Book Overview. What is this book about?

Chapter 2 — Attention Is Not Awareness

Why focus alone does not create clarity

We live in an age obsessed with focus.

We praise the ability to concentrate for long hours. We design tools to block distractions. We measure productivity by how long attention can be sustained on a task. We speak of "deep work," "flow," and "laser focus" as if attention itself were the pinnacle of human capability.

Yet despite all this effort, clarity remains elusive for many.

People are focused—and still overwhelmed.
Attentive—and still reactive.
Concentrating—and still emotionally blind.

This chapter explores a crucial distinction: **attention is not awareness**.
And without awareness, attention can become narrow, rigid, and misleading.

The Narrow Beam of Attention

Attention is a spotlight.
It selects.
It narrows.
It excludes.

When you focus on something, you amplify one stream of information by filtering out others. This is useful for tasks that require precision—reading a report, solving a problem, driving in traffic. Attention allows us to aim mental energy.

But attention has a limitation: it does not tell us how we are relating to what we are focused on.

You can be intensely attentive while:

- Being emotionally triggered
- Operating from fear or urgency
- Ignoring bodily signals
- Missing the wider context

In fact, the more threatened or pressured we feel, the *narrower* attention becomes.

Focus increases—but wisdom decreases.

Awareness: The Field, Not the Beam

Awareness is not a spotlight.
It is a field.

Where attention selects, awareness includes.
Where attention points, awareness surrounds.
Where attention locks on, awareness remains open.

Awareness notices:

- What you are focused on
- What you are feeling while focusing
- What is happening in your body
- What assumptions are shaping perception

- What is being excluded from view

Awareness does not compete with attention—it **governs it**.

Without awareness, attention becomes mechanical.
With awareness, attention becomes intelligent.

When Focus Becomes a Trap

Modern life trains attention relentlessly but neglects awareness entirely.

We are taught to:

- Push through discomfort
- Override fatigue
- Ignore emotional signals
- Stay "on task" at all costs

This creates a dangerous pattern: **high attention with low awareness**.

In this state:

- You may solve the wrong problem very efficiently
- You may escalate conflict while believing you are being rational
- You may miss ethical or emotional consequences
- You may feel busy but disconnected

Attention without awareness is like driving fast with fogged windows.

Speed increases. Visibility decreases.

Emotional Blind Spots

One of the most costly consequences of confusing attention with awareness is emotional blindness.

You can be fully focused on a conversation and still:

- Miss the tension in your voice
- Ignore defensiveness in your posture
- Overlook discomfort in the other person
- Fail to notice your own need for control

This is why people often say, "I was paying attention—why did this go wrong?"

Because attention heard the words.
Awareness would have felt the undercurrent.

Emotional intelligence begins not with better listening, but with **better awareness of the inner state that is listening**.

Awareness Creates Choice

Attention locks onto what *is*.
Awareness creates space around it.

That space is where choice lives.

Without awareness:

- Reaction feels inevitable
- Urgency feels justified

- Habits run unnoticed

With awareness:

- Reaction slows
- Perspective widens
- Response becomes possible

Awareness does not eliminate focus—it **humanizes it**.

It allows you to notice:

"I am focused—and I am tense."
"I am attentive—and I am afraid."
"I am concentrating—and I am narrowing."

That noticing changes everything.

Why Clarity Requires More Than Focus

Clarity is not produced by intensity.
It is produced by *orientation*.

You can stare harder at a situation and still not see it clearly.
You can analyze longer and still miss what matters.

Clarity emerges when:

- Attention is guided by awareness
- Focus is held within a wider field
- The inner state is as visible as the outer task

This is why presence—not focus—is the foundation of
emotional intelligence.

Presence includes attention.
But it also includes awareness of self, state, and context.

A Quiet Reframe

The problem is not that we lack focus.
It is that we trust focus too much.

We have mistaken intensity for insight.
Control for clarity.
Attention for awareness.

The work ahead is not to concentrate harder—but to **see wider**.

In the next chapter, we will explore what happens when awareness collapses under pressure—and how emotional fragmentation becomes the hidden cost of modern life.

Chapter 3 — The Illusion of Control

How reacting feels productive—but erodes stability

Control has a particular seduction.
It promises safety, certainty, and forward motion. It gives the nervous system a feeling of agency in moments of uncertainty. And in modern life—where speed is rewarded and decisiveness is praised—control often masquerades as competence.

But much of what we call control is not stability.
It is reaction.

Why Reaction Feels Like Progress

Reaction creates motion. Motion creates the sensation of doing something. And doing something—anything—can feel preferable to sitting with uncertainty. When pressure rises, reacting provides immediate relief. It tightens the reins. It sharpens tone. It accelerates decision-making. It looks decisive from the outside.

Internally, however, something else is happening.

Reaction narrows perception. It prioritizes speed over accuracy, certainty over clarity. The body moves into a defensive posture—muscles tighten, breath shortens, attention contracts. The mind scans for threats rather than possibilities.

In this state, action feels productive because it is fast.
But speed is not the same as effectiveness.

The Nervous System Behind Control

Control is often an attempt to regulate discomfort rather than circumstances. When the nervous system perceives threat—emotional, social, or cognitive—it seeks immediate ways to restore equilibrium. Commanding, correcting, interrupting, and deciding quickly are all strategies to reduce internal unease.

The problem is not the desire for regulation.
The problem is mistaking reaction for regulation.

True regulation expands capacity. Reaction consumes it.

Over time, habitual reactivity erodes stability rather than strengthening it. Decisions become brittle. Relationships become strained. Emotional energy is spent managing urgency instead of cultivating clarity.

The Hidden Cost of Staying "On Top of Things"

Many high-functioning individuals live under the belief that constant vigilance is responsible leadership. They stay alert, involved, and ready to intervene. They answer quickly. They solve immediately. They maintain control.

What goes unnoticed is the cost.

Constant reaction keeps the system in a low-grade state of threat. The body never fully returns to baseline. The mind never fully opens. Creativity diminishes. Listening becomes selective. Emotional signals—both one's own and others'—are filtered through urgency.

Stability is not created by staying on top of everything.
It is created by knowing when not to react.

Control Versus Presence

Control focuses on outcomes. Presence focuses on
conditions.

Control asks, How do I fix this now?
Presence asks, What is actually happening here?

Control seeks certainty. Presence tolerates ambiguity long
enough for insight to emerge. Control compresses time.
Presence expands it—often by just a few seconds, but enough
to change the quality of response.

This is where emotional intelligence begins to take shape. Not
in the absence of action, but in the space before it.

The Paradox of Stability

Stability does not come from tightening grip.
It comes from creating internal steadiness.

When the inner state is stable, external complexity becomes
more navigable. Decisions are slower but wiser.
Communication is firmer but calmer. Authority is felt without
being forced.

Presence does not remove challenges.
It removes the need to fight them immediately.

Relearning Control as Choice

The most powerful form of control is not reaction—it is choice.

Choice requires a pause.
A moment of awareness.
A brief return to the body.

In that moment, the nervous system shifts from defense to discernment. Options widen. Tone softens. The system regains flexibility.

This is not passivity. It is mastery.

A Quiet Reframe

Ask yourself:

- When pressure rises, do I react—or do I stabilize?
- Does my urgency create clarity, or does it silence it?
- Am I trying to control the situation, or regulate my state?

The illusion of control thrives on speed.
Presence restores stability through awareness.

And stability, once reclaimed, does not need to announce itself. It is felt—in decisions that land cleanly, in relationships that breathe, and in a leadership presence that steadies the room simply by being there.

Chapter 4 — Defining Presence

Beyond mindfulness, calmness, and concentration

Presence is a word that feels familiar—and yet remains strangely undefined.

It is often used interchangeably with *mindfulness*, *calmness*, *focus*, or *being in the moment*. While these states may accompany presence, they are not the same thing. Presence is not a technique you apply, a mood you maintain, or a skill you switch on. It is a *state of inner availability* - the condition that makes emotional intelligence possible at all.

To define presence clearly, we must first remove what it is *not*.

Presence Is Not Mindfulness (Though Mindfulness Can Support It)

Mindfulness is a practice.
Presence is a state.

Mindfulness trains attention—often through structured exercises such as observing breath, sensation, or thought without judgment. These practices can *strengthen* the capacity for presence, but presence itself does not require formal mindfulness techniques.

You can be mindful and still unavailable.
You can observe thoughts and still be defended.
You can notice sensations and still be emotionally absent.

Presence begins not with observation, but with *openness*.

It is possible to be deeply present while engaged in action, conversation, conflict, leadership, or decision-making—without silently watching your breath or narrating your experience. Presence is not inward withdrawal; it is inward *grounding*.

Presence Is Not Calmness

Calmness is a nervous system state.
Presence is a relational one.

Many people assume presence means being relaxed, unruffled, or emotionally smooth. This misunderstanding creates pressure to *appear* calm—often leading to emotional suppression.

Presence can coexist with:

- Intensity
- Uncertainty
- Strong emotion
- Pressure
- Discomfort

A leader can be present while anxious.
A parent can be present while overwhelmed.
A conversation can be present while charged.

Calmness may emerge from presence—but it is not a prerequisite for it.

In fact, the attempt to *force calm* often pulls us out of presence, because it shifts attention toward control rather than contact.

Presence Is Not Concentration

Concentration narrows attention.
Presence widens awareness.

Concentration focuses on *one thing* at the exclusion of others. Presence allows multiple layers of information to be held simultaneously:

- Internal sensations
- Emotional signals
- Environmental cues
- Relational dynamics

You can concentrate intensely while being disconnected from your body, emotions, or the people around you. Presence, by contrast, keeps you *available to what is happening now—* without collapse or defense.

Presence is not tunnel vision.
It is panoramic awareness.

A Working Definition of Presence

Presence is the capacity to remain internally available to reality—without rushing to escape, control, or react.

It is the ability to:

- Stay with experience as it is
- Sense before responding
- Allow emotion without being overtaken by it
- Engage without being hijacked

Presence is not passive.
It is not detached.
It is not slow.

It is *stable*.

Why Presence Is the Foundation of Emotional Intelligence

Emotional intelligence requires several abilities:

- Noticing emotional signals
- Regulating reactions
- Understanding others
- Making considered choices

All of these depend on one condition: the system must be present enough to receive information.

Without presence:

- Emotions are felt only after they erupt
- Reactions precede awareness
- Empathy becomes projection
- Self-regulation becomes suppression

Presence is the platform on which every EQ skill operates.

You cannot regulate what you cannot sense.
You cannot choose what you cannot see.
You cannot lead from a state you are not aware of.

Presence as Inner Posture

Presence is best understood not as a behavior, but as an *inner posture*.

An inner posture of:

- Receptivity rather than resistance
- Curiosity rather than judgment
- Stability rather than urgency

This posture allows life to be met directly, rather than through habitual filters of fear, control, or performance.

Presence does not remove difficulty.
It removes distortion.

The Cost of Undefined Presence

When presence is poorly defined, people chase its substitutes:

- Calm instead of clarity
- Focus instead of awareness
- Techniques instead of states

This leads to frustration:
"I meditate, but I still react."

"I'm focused, but disconnected."
"I'm calm, but not clear."

The missing element is not effort.
It is presence.

Presence Is Always Available—But Rarely Occupied

Presence does not need to be created.
It needs to be *returned to*.

It is available:

- In conversation
- In tension
- In decision-making
- In conflict
- In silence

But it requires one essential condition: the willingness to pause long enough to arrive.

In the next chapter, we will explore what happens in the body and nervous system when presence is available—and why modern life makes this state increasingly difficult to access without intention.

Presence is not something you do.
It is where you come from.

And everything that follows - emotionally, relationally, and ethically - depends on that place.

Chapter 5 — Presence and the Nervous System

The physiology beneath emotional behavior

Presence is not an abstract virtue.
It is a *physiological condition*.

Every emotional response, every reaction, every moment of clarity or collapse is mediated through the nervous system. Long before thought forms, before words appear, before choice becomes conscious, the body has already assessed the situation and shifted its internal state.

To understand presence, we must understand the system that makes it possible—or impossible.

The Nervous System as the Hidden Operator

The nervous system is constantly asking one primary question:

"Am I safe?"

This assessment happens automatically, continuously, and below conscious awareness. It scans:

- Tone of voice
- Facial expression
- Speed and unpredictability
- Power dynamics
- Memory and association
- Internal sensations

Based on this assessment, the system shifts into one of two broad orientations:

- Regulation (safety)
- Protection (threat)

Presence lives only in the first.

Regulation vs. Protection

When the nervous system perceives sufficient safety, it enters a regulated state. In this state:

- Attention widens
- Breathing deepens
- Emotional signals are accessible
- The body remains responsive, not reactive

This is the physiological ground of presence.

When the system detects threat—real or perceived—it shifts into protection:

- Fight (irritation, anger, control)
- Flight (avoidance, anxiety, distraction)
- Freeze (shutdown, numbness, collapse)

In protection, presence disappears—not because of failure, but because the system is doing its job.

Why Emotional Intelligence Fails Under Threat

Emotional intelligence is often treated as a cognitive skill -
something we *apply*.

But cognition is only available when the nervous system allows
it.

Under threat:

- The body prioritizes survival over reflection
- Reaction precedes awareness
- Emotion becomes force rather than information
- Empathy narrows or vanishes

This is why people say things they later regret.
Why intelligent leaders behave irrationally under pressure.
Why insight disappears in moments of conflict.

It is not a character flaw.
It is physiology.

Presence Is a Regulated State, Not a Moral One

Presence is not about being "better," "calmer," or "more
evolved."

It is about whether the nervous system is regulated enough to
allow *choice*.

A person in protection cannot:

- Pause reliably
- Reflect accurately
- Access empathy consistently

- Respond proportionally

Asking someone to "be more present" without addressing nervous system state is like asking a drowning person to swim more gracefully.

Presence is not willed.
It is *supported*.

The Myth of Control

Many people attempt to manage emotional behavior through willpower:

- "I should stay calm."
- "I must not react."
- "I need to control my emotions."

This approach ignores the body.

Control strategies often *increase* threat, because they introduce self-judgment and internal pressure. The nervous system reads this as additional danger.

Presence, by contrast, reduces threat through *permission*:

- Permission to feel
- Permission to slow
- Permission to sense

This signals safety—not perfection.

The Body as the Gateway to Presence

Presence does not begin in thought.
It begins in sensation.

Before you can name an emotion, the body has already
registered:

- Tightness
- Heat
- Constriction
- Expansion
- Movement

When these sensations are allowed—without immediate
interpretation or resistance—the nervous system stabilizes.

This is why presence feels grounding.
Not because problems disappear, but because the body is no
longer bracing against itself.

Why Modern Life Disrupts Regulation

Modern environments continuously activate protection:

- Speed and urgency
- Constant evaluation
- Digital overstimulation
- Chronic comparison
- Lack of recovery

The nervous system was not designed for sustained alertness
without resolution.

As a result, many people live in a *baseline of mild threat*:

- Functional, but tense
- Productive, but reactive
- Connected, but not available

Presence becomes rare—not because it is difficult, but because regulation is unsupported.

Presence as Nervous System Literacy

Developing presence is not about mastering techniques.
It is about learning to *read* and *respect* the nervous system.

This includes:

- Recognizing early signs of activation
- Understanding personal threat patterns
- Knowing when regulation is needed before engagement
- Allowing the body to settle before demanding clarity

Presence grows as safety increases.

Emotional Behavior Is a State Issue, Not a Trait Issue

People often label themselves—or others—as:

- "Emotionally reactive"
- "Too sensitive"
- "Cold"
- "Detached"
- "Overwhelmed"

Most of these are not traits.
They are *states*.

Change the state of the nervous system, and behavior changes
naturally.
Not through force—but through access.

Presence Restores Choice

When the nervous system is regulated:

- Emotions become signals, not commands
- Thought regains flexibility
- Behavior aligns with values rather than impulse

This is the true power of presence.

Not calmness.
Not control.
Not suppression.

But choice.

In the next chapter, we will explore how presence interrupts
reactive loops—and how small moments of awareness can
prevent emotional escalation before it begins.

Presence is not an idea.
It is a biological invitation.

And when the body accepts it, emotional intelligence becomes
possible again.

Chapter 6 — Why Presence Collapses Under Pressure

Stress, identity, fear, and emotional narrowing

Presence does not disappear because we forget its importance.
It collapses because pressure changes how the nervous system prioritizes survival.

Under pressure, the system does not ask, "What is the wisest response?"
It asks, "What protects me right now?"

This shift is subtle, automatic, and deeply human. To understand why presence collapses, we must look at how stress, identity, and fear narrow inner space—and how emotional intelligence becomes unavailable when self-protection takes the wheel.

Pressure Is Not the Problem—Interpretation Is

Pressure itself is not inherently harmful. Deadlines, responsibility, and challenge can sharpen engagement. Presence collapses when pressure is *interpreted* as threat.

The nervous system reads threat not only from danger, but from:

- Time scarcity
- High stakes
- Evaluation and judgment

- Loss of status or control
- Emotional unpredictability

When these cues accumulate, the system shifts from *engagement* to *defense*.

This is the moment presence contracts.

Stress Narrows the Field of Awareness

Under stress, awareness shrinks.

What was once a wide field becomes a narrow tunnel:

- Fewer options appear
- Nuance disappears
- Emotional range collapses
- Binary thinking dominates

This is emotional narrowing.

It explains why intelligent people:

- Miss obvious alternatives
- Overreact to small triggers
- Speak with unnecessary force
- Become rigid or withdrawn

The system is not failing.
It is protecting.

Identity Enters the Equation

Pressure becomes exponentially stronger when it attaches to identity.

Identity answers the question: *"What does this mean about me?"*

Stress escalates when situations threaten:

- Competence
- Worth
- Authority
- Belonging
- Image

At this point, the nervous system is no longer responding to the present moment. It is responding to *who we think we must be* in that moment.

Presence collapses because the system is no longer available to reality—it is busy defending identity.

Fear Speaks Faster Than Thought

Fear does not wait for language.

It shows up as:

- Urgency
- Tightness
- Irritation
- Defensiveness
- Withdrawal

These are not emotions chosen consciously.
They are survival signals.

Fear narrows attention because broad awareness requires safety. When fear enters, the system prioritizes speed over accuracy, certainty over complexity.

This is why pressure creates:

- Reactive communication
- Control behavior
- Micromanagement
- Emotional withdrawal
- Loss of empathy

Presence cannot coexist with fear that feels unmanaged.

Emotional Intelligence Requires Space—Pressure Removes It

Every EQ skill depends on *space*:

- Space to notice emotion
- Space to pause
- Space to consider impact
- Space to choose response

Pressure compresses that space.

Under compression:

- Awareness is delayed
- Regulation is replaced by reaction

- Empathy becomes self-reference
- Listening turns into preparing rebuttals

Presence collapses not because people lack skill—but because the inner space required for skill has vanished.

The Body Braces Before the Mind Knows

Before conscious awareness registers pressure, the body has already reacted:

- Breath becomes shallow
- Muscles tighten
- Jaw clenches
- Shoulders rise
- Heart rate shifts

These are not symptoms to correct.
They are signals to listen to.

When these signals are ignored, pressure escalates silently—until presence disappears entirely.

Why "Trying Harder" Makes It Worse

When presence collapses, many people respond with effort:

- More control
- More force
- More self-criticism
- More urgency

This compounds the problem.

Effort without regulation signals danger to the nervous system. The system interprets internal pressure as additional threat, further narrowing awareness.

Presence does not return through force.
It returns through *safety*.

Pressure Turns Values Into Performance

Under pressure, values are often replaced by performance:

- Listening becomes convincing
- Leadership becomes dominance
- Care becomes efficiency
- Integrity becomes image management

This shift is rarely intentional.
It is a by-product of fear.

Presence collapses when being *right*, *respected*, or *in control* feels more urgent than being *real*, *connected*, or *aware*.

The Hidden Cost of Chronic Pressure

When pressure is constant, collapse becomes the baseline.

People describe this as:

- "I'm functioning, but not myself."
- "I'm productive, but disconnected."

- "I don't know why I'm so reactive."

This is not burnout yet—but it is erosion.

Presence cannot be sustained in a system that never returns to safety.

Presence Returns When Threat Is Reduced, Not Eliminated

Presence does not require the absence of pressure.
It requires the reduction of *perceived threat*.

This happens through:

- Slowing internal pace
- Naming what is happening
- Allowing sensation without resistance
- Separating identity from outcome

These are not techniques.
They are conditions.

From Collapse to Capacity

The collapse of presence under pressure is not a weakness.
It is information.

It shows us where safety is lacking.
Where identity is entangled.
Where fear is driving behavior.

In the next chapter, we will explore how presence can be rebuilt *within* pressure—without waiting for calm, certainty, or ideal conditions.

Presence does not disappear because we are incapable.
It disappears because we are human.

And when we understand why it collapses, we gain the power to restore it—one moment of awareness at a time.

Chapter 7 — Self-Awareness Begins With Presence

Seeing clearly without judgment

Self-awareness is often described as knowing your emotions, understanding your patterns, or recognizing your triggers. While these descriptions are not wrong, they skip the most essential condition that makes self-awareness possible in the first place.

Self-awareness does not begin with analysis.
It begins with presence.

Without presence, what we call self-awareness is usually hindsight—interpretation after the fact, explanation layered over reaction. We notice what happened only once the emotion has already driven the behavior. True self-awareness happens earlier, quieter, and closer to the moment of experience itself.

It happens when awareness is present *before* judgment arrives.

The Mistake We Make About Self-Awareness

Many people approach self-awareness as a cognitive skill. They try to think their way into understanding themselves:

- Why did I react like that?
- What does this say about me?
- Is this fear, anger, or insecurity?

While reflection has value, it often comes too late to change the emotional trajectory. By the time the mind is asking these questions, the nervous system has already moved. The emotion has already shaped tone, posture, words, and decisions.

Presence works differently.

Presence does not ask *why* first.
It notices *what is*.

A tightening in the chest.
A rush of heat.
A contraction behind the eyes.
A subtle urge to defend, explain, withdraw, or dominate.

This noticing happens without commentary. Without story. Without blame.

And that is what makes it powerful.

Awareness Without Judgment Is the Turning Point

Judgment collapses awareness.

The moment we label an emotion as *bad*, *wrong*, *weak*, or *unacceptable*, the system moves into self-protection. Awareness narrows. Curiosity disappears. Defense replaces observation.

Presence creates a different internal environment.

In presence, emotions are allowed to exist without needing to be fixed, justified, or acted upon. This does not make them stronger. It makes them *visible*.

Visibility changes everything.

When an emotion is seen clearly, without resistance, it no longer needs to escalate to be noticed. Much of emotional intensity is not about the feeling itself, but about the system trying to push the feeling into awareness.

Presence removes the need for that escalation.

The Observer Position

Presence establishes an inner vantage point—a subtle but profound shift from *being inside the emotion* to *being aware of the emotion*.

This is not detachment.
It is not suppression.
It is not distancing.

It is proximity without fusion.

From this observer position:

- Emotions are experienced fully, but not blindly
- Thoughts are noticed, but not believed automatically
- Urges are felt, but not immediately obeyed

This is the birthplace of emotional intelligence.

Not control.
Not positivity.
Not calm at all costs.

But clarity.

Why Presence Makes Honesty Possible

Most people believe they lack self-awareness because they are not honest enough with themselves. In reality, they lack the *capacity to stay present* with what they feel.

Honesty requires safety.

When the inner environment is harsh, critical, or impatient, awareness avoids depth. It skims the surface. It edits. It rationalizes.

Presence creates psychological safety internally.

When there is no immediate judgment, the system relaxes. It allows uncomfortable truths to surface:

- Fear beneath confidence
- Grief beneath anger
- Insecurity beneath control
- Fatigue beneath irritability

These layers are not discovered through force. They reveal themselves when awareness is steady and non-threatening.

Seeing Clearly Changes Choice

Self-awareness is not an end in itself.
Its value lies in what it makes possible.

When you see clearly:

- You recognize emotional momentum before it carries you away
- You notice patterns before they repeat automatically
- You sense when a response is being driven by fear rather than intention

Presence creates a pause—not as a technique, but as a natural consequence of clarity.

And within that pause, choice re-enters.

Not perfect choice.
Not always wise choice.

But *conscious* choice.

Presence Is the Foundation, Not the Outcome

Many people try to become present *after* becoming self-aware.
The truth is the reverse.

Presence is not the reward for understanding yourself.
It is the ground that makes understanding possible.

Without presence:

- Self-awareness becomes self-criticism
- Reflection becomes rumination
- Insight becomes identity reinforcement

With presence:

- Awareness becomes spacious
- Insight becomes gentle
- Growth becomes sustainable

A Quiet Reframing

Self-awareness does not mean constantly monitoring yourself.
It does not mean fixing every emotional response.
It does not mean becoming endlessly introspective.

It means being *here* enough to notice what is happening before judgment hardens it into identity.

Presence is not something you add to self-awareness.

Presence *is* self-awareness—before it turns into thought.

And from that place, seeing clearly becomes natural.

Not because you try harder,
but because you stop looking away.

Chapter 8 — Regulation Without Suppression

Stability without emotional shutdown

Many people believe regulation means control.

Control over tone.
Control over reaction.
Control over what is shown and what is hidden.

In practice, this understanding of regulation often produces the opposite of what is intended. Instead of stability, it creates tension. Instead of clarity, it creates rigidity. Instead of emotional intelligence, it produces emotional numbness or delayed overwhelm.

True regulation is not suppression.
It is **capacity**.

It is the ability to remain present with emotional energy without being flooded by it—and without needing to shut it down.

This chapter explores what regulation actually is, why suppression is so commonly mistaken for maturity, and how presence allows stability to emerge naturally, without emotional shutdown.

The Common Misunderstanding of Regulation

From an early age, many of us learn that strong emotions are a problem to be managed.

"Calm down."
"Don't overreact."
"Be professional."
"Stay composed."

These instructions are not inherently harmful. But when they are not paired with inner awareness, they teach a subtle lesson:
emotions must be contained, hidden, or overridden.

Over time, this leads to a form of pseudo-regulation—an external appearance of calm built on internal disconnection.

People who suppress emotions often appear:

- Controlled
- Rational
- Unaffected
- High-functioning

But beneath that surface, emotional energy does not disappear. It accumulates.

Suppression does not reduce emotion.
It postpones it.

And postponed emotion eventually emerges as:

- Sudden outbursts
- Chronic fatigue
- Irritability
- Emotional numbness
- Loss of empathy
- Physical tension or illness

What looks like regulation from the outside is often **shutdown on the inside**.

Suppression Narrows the Inner Field

Suppression requires effort.

To suppress an emotion, the system must actively push against it—tightening muscles, restricting breath, narrowing attention, and disengaging from sensation.

This narrowing has consequences.

When the inner field contracts:

- Awareness reduces
- Sensitivity dulls
- Choice diminishes
- Reactivity moves underground

The person may seem calm, but the nervous system remains activated. The energy that could have been processed is now trapped.

This is why suppressed individuals often feel "fine" until they suddenly aren't.

Regulation that depends on suppression is unstable by design.

What Regulation Actually Is

True regulation is not about removing emotion.
It is about **staying connected while emotions move**.

Regulation means:

- The nervous system can experience activation without tipping into overwhelm
- Sensation is allowed without needing immediate action
- Emotion is felt without being acted out or pushed away
- Awareness remains wider than the emotion itself

In regulation, emotions are neither indulged nor resisted.
They are **held within presence**.

This holding is not mental.
It is physiological and attentional.

The body remains open.
The breath remains available.
Awareness stays spacious.

From this state, emotion completes its cycle instead of being interrupted.

Presence Is the Missing Mechanism

Presence is what makes regulation possible without force.

When awareness is present:

- Sensations are noticed early
- Intensity is met gradually
- The nervous system receives signals of safety

- Emotional energy is allowed to move through

Presence prevents escalation not by stopping emotion, but by **preventing contraction**.

Without presence, people attempt regulation at the level of behavior:

- Don't say that
- Don't show this
- Don't feel so much

With presence, regulation occurs at the level of state:

- The body stays grounded
- The mind stays open
- Emotion moves without hijacking action

This is the difference between control and stability.

Stability Is Not Flatness

One of the quiet costs of suppression is emotional flattening.

People often mistake this flatness for maturity.

But emotional intelligence does not eliminate depth—it preserves it.

Healthy regulation allows:

- Sadness without collapse
- Anger without aggression
- Fear without paralysis

- Joy without loss of grounding

Stability does not mean less feeling.
It means **more capacity for feeling**.

A regulated system can experience intensity without becoming distorted by it.

Why Shutdown Feels Safer Than Presence

For many, shutdown feels safer than openness.

Presence requires staying with sensation.
Shutdown avoids it.

Especially for those who have experienced chronic stress, conflict, or emotional unpredictability, suppression becomes a survival strategy.

It works—temporarily.

But survival strategies are not growth strategies.

What once protected you eventually limits you.

Presence gently retrains the system to recognize that emotion is not inherently dangerous—and that awareness is sufficient to meet it.

Regulation Is a Relationship, Not a Technique

There is no single technique that creates regulation.

Regulation emerges from a relationship with inner experience.

A relationship marked by:

- Curiosity instead of judgment
- Allowing instead of resisting
- Attention instead of avoidance

This relationship is built moment by moment, not through effort, but through orientation.

You do not regulate emotion by fixing it.
You regulate emotion by **staying with it without collapsing or constricting**.

The Quiet Strength of Non-Shutdown

The most stable people are not the least emotional.

They are the least defended.

They do not need to suppress, because they are not afraid of what arises.

Their presence provides containment.
Their awareness provides space.
Their nervous system knows how to move through activation without losing coherence.

This is not stoicism.
It is not detachment.
It is not emotional restraint.

It is embodied steadiness.

What This Chapter Leaves You With

Regulation without suppression is not about doing something new.
It is about **stopping what disrupts natural regulation**.

When presence replaces resistance:

- The body recalibrates
- Emotions complete their cycle
- Stability becomes sustainable

You do not need to shut down to stay steady.
You need enough presence to remain open.

In the next chapter, we explore how this inner steadiness transforms communication—allowing expression without reactivity, and honesty without harm.

Chapter 9 — Empathy Without Self-Loss

Staying open without emotional exhaustion

Empathy is often celebrated as an unquestioned virtue.

To care deeply.
To feel with others.
To be emotionally available.

Yet for many people, empathy becomes a source of depletion rather than connection. What begins as openness quietly turns into fatigue. What starts as care slowly becomes burden. Over time, empathy collapses into withdrawal, irritability, or numbness.

This chapter explores why empathy so often leads to exhaustion—and how presence allows empathy to remain spacious, grounded, and sustainable, without self-loss.

When Empathy Becomes Drain

Empathy itself is not the problem.

The problem is how empathy is held.

Many people confuse empathy with emotional merging. They unconsciously cross an inner boundary where another person's emotional state becomes their own. Instead of witnessing emotion, they absorb it. Instead of staying present, they become entangled.

This form of empathy feels like:

- Carrying emotions that are not yours
- Feeling responsible for how others feel
- Losing clarity when someone else is distressed
- Needing distance to recover afterward
- Emotional exhaustion after "being there" for others

What is called compassion fatigue is often not too much compassion—but **too little presence**.

Empathy Without Presence Is Porous

Without presence, empathy lacks structure.

When awareness is narrow and ungrounded:

- Emotions move unchecked across internal boundaries
- The nervous system mirrors others without regulation
- Identity blurs with relational roles
- Emotional energy leaks without recovery

This is why highly empathetic people are often overwhelmed.

Their sensitivity is real.
Their capacity to hold it is not yet stable.

Presence provides the container empathy requires.

The Difference Between Resonance and Absorption

Healthy empathy involves **resonance**, not absorption.

Resonance means:

- You sense another's emotional state
- You remain anchored in your own body
- Awareness stays wider than the emotion
- Feeling arises without ownership

Absorption means:

- You take on the emotion as your own
- The body tightens or drains
- Perspective narrows
- Recovery requires withdrawal

The distinction is subtle—but critical.

Empathy with presence allows connection **without collapse**.

Presence Restores Inner Boundaries

Boundaries are often misunderstood as walls.

True boundaries are not rigid barriers.
They are **clarity of self-location**.

Presence restores this clarity.

When you are present:

- You know where your experience ends and another's begins
- Sensation is felt without confusion
- Compassion flows without entanglement
- You can care without carrying

Boundaries emerge naturally when awareness is embodied.

They do not need to be enforced.
They are sensed.

Why Empathy Exhausts the Nervous System

Emotional exhaustion is a physiological event.

When the nervous system remains in prolonged empathic activation without grounding, it does not distinguish between external and internal threat. Emotional mirroring triggers the same stress responses as direct experience.

Without regulation:

- The body stays mobilized
- Cortisol remains elevated
- Emotional energy is consumed
- Recovery becomes slow or incomplete

Presence interrupts this cycle.

By staying connected to your own sensations—breath, posture, contact with the ground—the nervous system remains oriented and stable while empathy occurs.

Compassion Is Spacious, Not Heavy

There is a quiet misunderstanding that compassion must hurt.

That to care deeply means to feel deeply wounded.

But true compassion has **space**.

It includes warmth without urgency.
Concern without collapse.
Care without self-sacrifice.

When presence is established, empathy does not require emotional pain as proof of sincerity.

You can be fully human without being depleted.

Self-Loss Is Not a Requirement for Connection

Many people unconsciously believe:
"If I don't feel what they feel, I'm being cold."
"If I don't carry this, I'm being selfish."
"If I stay steady, I'm not caring enough."

These beliefs confuse empathy with self-erasure.

But losing yourself does not help others.

In fact, self-loss reduces your ability to respond wisely, to listen clearly, and to remain available over time.

Presence allows empathy without abandonment of self.

The Strength of Grounded Empathy

Grounded empathy looks different.

It is quieter.
Less dramatic.
More stable.

It shows up as:

- Listening without urgency to fix
- Feeling with someone without becoming overwhelmed
- Remaining calm in emotional intensity
- Offering steadiness instead of absorption

This form of empathy does not drain.
It **regulates the space**.

Others often feel safer, not because you carry their emotions—
but because you are not destabilized by them.

Staying Open Without Burning Out

Sustainable empathy rests on three inner orientations:

- Awareness stays wider than emotion
- The body remains anchored
- Care flows without ownership

These are not techniques.
They are qualities of presence.

When presence is strong:

- Openness no longer equals vulnerability to depletion
- Compassion does not require collapse
- Emotional availability becomes renewable

You can remain open without being emptied.

What This Chapter Leaves You With

Empathy is not meant to cost you yourself.

When presence leads:

- Connection deepens
- Boundaries clarify
- Exhaustion recedes
- Compassion stabilizes

You do not need to harden to protect yourself.
You do not need to withdraw to recover.

You need enough presence to stay **with** others—without leaving yourself behind.

In the next chapter, we explore how presence transforms leadership—not as authority or control, but as emotional gravity that others naturally orient toward.

Chapter 10 — Leading From Presence

Authority, calm, and credibility under pressure

Leadership is often described in terms of vision, strategy, and decision-making.

But long before people evaluate what a leader says, they respond to **how the leader is**.

Under pressure, leadership is felt more than heard.

This chapter explores why presence—not personality, position, or performance—is the invisible source of authority, calm, and credibility, especially when conditions are uncertain or emotionally charged.

Leadership Is a State, Not a Role

Many people step into leadership roles without stepping into leadership states.

They have responsibility, but not regulation.
They have authority, but not steadiness.
They have visibility, but not presence.

When pressure rises, this gap becomes visible.

People do not look to leaders for perfection.
They look for **orientation**.

They ask, often unconsciously:

- Is this person grounded?

- Can they tolerate uncertainty?
- Do they react—or respond?
- Do they stabilize the room, or amplify tension?

Leadership begins before action—at the level of state.

Why Calm Is Interpreted as Competence

In moments of uncertainty, the nervous systems of others seek cues of safety.

Leaders provide those cues constantly.

When a leader is reactive:

- Urgency spreads
- Fear accelerates
- Clarity collapses

When a leader is present:

- Pace slows
- Attention widens
- Thinking improves
- Confidence stabilizes

Calm is not a personality trait.
It is a physiological signal.

Presence communicates, "We are not lost—even if we don't yet know the way."

Authority Without Force

True authority does not need to assert itself.

It does not rely on volume, urgency, or control.
It emerges from **coherence**.

When inner experience is regulated:

- Words land with weight
- Decisions feel grounded
- Boundaries are respected
- Direction is trusted

This form of authority is felt, not demanded.

People align not because they are told to—but because the leader's presence provides stability they can orient toward.

Pressure Reveals, Not Creates, Leadership

Pressure does not turn someone into a leader.

It reveals the inner structures that already exist.

Under stress:

- Unintegrated fear surfaces
- Ego seeks control
- Certainty is performed
- Listening narrows

Presence changes this equation.

A present leader can:

- Admit uncertainty without losing credibility
- Hold tension without rushing to resolve it
- Listen without defensiveness
- Decide without urgency

This capacity cannot be faked under pressure.

Credibility Is Emotional Before It Is Intellectual

Credibility is often assumed to come from expertise.

But in volatile conditions, expertise alone is not enough.

People assess credibility through emotional signals:

- Does this person seem grounded?
- Are they centered or scattered?
- Are they reacting or sensing?
- Do they create clarity or confusion?

Presence aligns internal state with external action.

When words, tone, posture, and timing are coherent, trust increases—even when answers are incomplete.

Leading Without Emotional Contagion

One of the most underestimated leadership skills is the ability to **not transmit stress**.

Leaders are emotional amplifiers.

Whatever they carry spreads:

- Anxiety
- Confidence
- Reactivity
- Stability

Presence interrupts emotional contagion.

A leader who is present does not absorb the group's anxiety—nor project their own.

Instead, they regulate the space.

Their steadiness allows others to settle, think, and respond more intelligently.

Presence Creates Psychological Safety

Psychological safety is not created through policies or statements.

It is created through moment-to-moment interaction.

When a leader is present:

- People speak more honestly
- Mistakes are addressed without shame
- Feedback is received without threat
- Conflict becomes workable

Presence signals that emotions can exist without punishment—and that truth does not require protection.

This is the soil in which performance and trust grow.

The Quiet Power of Not Knowing

One of the most powerful expressions of presence in leadership is the capacity to say:

"I don't know—yet."

Without collapse.
Without defensiveness.
Without urgency to fill the space.

This does not weaken authority.
It strengthens it.

Because it demonstrates:

- Confidence without arrogance
- Humility without insecurity
- Openness without loss of direction

Presence allows uncertainty to be held without panic.

Leadership That Endures Pressure

Leadership based on image eventually fractures.

Leadership based on presence **endures**.

Because presence:

- Regulates the nervous system
- Stabilizes attention
- Grounds decision-making

- Preserves relational trust

Under sustained pressure, people do not follow the most charismatic voice.

They follow the most stable center.

What This Chapter Leaves You With

Leading from presence does not require a new style.

It requires a new **orientation**.

When presence leads:

- Authority emerges naturally
- Calm becomes contagious
- Credibility remains intact—even in uncertainty

You do not need to perform leadership.

You need to **be anchored enough** that others can orient themselves through you.

In the next section of this book, we turn inward—exploring how presence is cultivated daily, not as a technique, but as a way of living that reshapes identity, resilience, and choice.

Chapter 11 — Presence in Difficult Conversations

Remaining grounded when emotions escalate

Most conversations are easy when emotions are low.

It is when voices tighten, defenses rise, and meaning becomes charged that presence is tested.

Difficult conversations do not fail because people lack intelligence or good intentions. They fail because emotional escalation narrows awareness, disrupts regulation, and pulls participants out of presence.

This chapter explores how presence allows you to remain grounded when conversations become charged—without withdrawing, dominating, or losing yourself.

Why Conversations Escalate

Escalation rarely begins with words.

It begins with **state**.

Before tone sharpens or positions harden, something shifts internally:

- A threat is sensed
- An identity feels challenged
- A boundary feels crossed
- A past wound is activated

The nervous system moves into protection.

Attention narrows.
Listening decreases.
The need to defend increases.

At this point, the conversation is no longer about the topic—it is about **safety**.

Presence Interrupts the Escalation Loop

Presence does not prevent emotion from arising.

It prevents emotion from **taking over**.

When presence is available:

- Sensations are noticed early
- Reactivity is recognized before it drives speech
- Awareness stays wider than the emotional surge
- Choice remains accessible

This creates a pause—not as a technique, but as a natural consequence of awareness.

Escalation requires unconsciousness.
Presence breaks the loop.

Staying in the Body While Speaking

One of the first things lost in difficult conversations is bodily awareness.

People leave their bodies and live in their arguments.

Presence restores embodiment.

Remaining grounded means:

- Feeling your feet or seat
- Allowing breath to remain available
- Noticing tension without trying to remove it
- Letting sensation anchor attention

This bodily grounding stabilizes the nervous system and prevents verbal overreaction.

You do not need to calm down before speaking.
You need to stay **connected** while speaking.

Listening Without Collapse or Defense

Listening under pressure is not passive.

It is an active regulation skill.

Presence allows you to listen without:

- Absorbing another's emotion
- Becoming defensive
- Planning your response prematurely
- Losing your own perspective

When you listen from presence:

- You hear both content and emotional signal
- You stay anchored in your own experience

- You do not confuse understanding with agreement

This form of listening lowers intensity—not because you comply, but because you are not resisting.

Speaking From State, Not Reaction

Words spoken from reactivity escalate.

Words spoken from presence stabilize—even when they are firm.

Presence shifts how speech lands:

- Tone becomes measured
- Timing improves
- Boundaries are expressed clearly
- Defensiveness reduces

You can say difficult things without aggression when your state is regulated.

It is not the words that escalate conflict.
It is the **state beneath them**.

When Emotions Rise in the Other Person

Presence is especially tested when the other person becomes emotional.

Raised voice.
Tears.

Accusation.

Withdrawal.

The instinct is to either:

- Match the intensity
- Fix the emotion
- Shut it down
- Escape the conversation

Presence offers another option: **stay steady**.

By remaining grounded:

- You do not mirror escalation
- You do not absorb distress
- You do not rush to resolve
- You allow emotion to move without amplifying it

Often, the other person settles—not because they are corrected, but because the space no longer feeds escalation.

Boundaries Without Emotional Armor

Difficult conversations often require boundaries.

Presence allows boundaries to be expressed without:

- Harshness
- Justification
- Emotional armor
- Withdrawal

A boundary spoken from presence is clear and calm.

It does not attack.
It does not apologize for existing.
It does not invite debate.

It simply states reality.

This kind of boundary often ends arguments rather than
escalating them.

The Power of Slowing Down

Speed escalates conflict.

Presence naturally slows pace.

This may look like:

- Pausing before responding
- Allowing silence
- Speaking fewer words
- Letting intensity pass before continuing

Slowness is not avoidance.

It is regulation in action.

When pace slows, the nervous system has space to
recalibrate—and the conversation can return to meaning
instead of momentum.

Repair When Presence Is Lost

No one remains present at all times.

What matters is the ability to **repair**.

Presence allows you to notice:

- "I reacted."
- "That landed harshly."
- "I need to reset."

Repair restores trust faster than perfection ever could.

A simple acknowledgment from presence often dissolves tension more effectively than continued argument.

What This Chapter Leaves You With

Difficult conversations are not problems to solve.

They are moments that require **capacity**.

When presence is available:

- Grounding replaces urgency
- Listening replaces defense
- Boundaries replace escalation

You do not need to dominate or withdraw when emotions rise.

You can remain **rooted**, responsive, and clear—even in the heat of the moment.

In the next chapter, we explore how presence reshapes identity itself—loosening rigid self-concepts and creating freedom from emotional roles that no longer serve you.

Chapter 12 — Decision-Making in Uncertainty

Clarity when answers are not obvious

Most decisions are not made in calm conditions.

They are made amid incomplete information, competing priorities, emotional pressure, and the quiet fear of getting it wrong.

Uncertainty is not the exception to decision-making.
It is the environment in which most meaningful decisions occur.

This chapter explores how presence creates clarity when answers are not obvious—without rushing, overthinking, or freezing.

Why Uncertainty Feels So Uncomfortable

Uncertainty activates a primal response.

The nervous system seeks closure.
The mind searches for certainty.
Emotion pushes toward action.

Not because action is wise—but because **not knowing feels unsafe**.

When uncertainty rises:

- The mind fills gaps with assumptions

- The need for control intensifies
- Speed replaces discernment
- Decisions are made to reduce discomfort rather than increase accuracy

Many poor decisions are not the result of bad thinking.
They are the result of **unregulated urgency**.

Presence Separates Urgency from Importance

Presence introduces a critical distinction.

Not everything that feels urgent is important.
Not everything that feels uncomfortable requires immediate action.

When awareness is present:

- Sensations of pressure are noticed without being obeyed
- Emotional signals are distinguished from factual signals
- Time expands just enough for discernment

Presence does not eliminate pressure.
It prevents pressure from hijacking choice.

Clarity Is Not Certainty

A common misunderstanding is that clarity means knowing the outcome.

It does not.

Clarity means:

- Knowing what you are responding to
- Understanding what matters most right now
- Recognizing what you know—and what you do not
- Sensing when more information is needed
- Knowing when enough is enough

Presence allows clarity **without guarantees**.

You may still choose incorrectly.
But you choose **consciously**.

When the Mind Overcompensates

In uncertainty, the mind often tries to compensate through analysis.

More data.
More scenarios.
More opinions.
More thinking.

While analysis is valuable, it becomes counterproductive when driven by anxiety rather than inquiry.

Signs of overcompensation include:

- Endless comparison without resolution
- Replaying the same options repeatedly
- Seeking reassurance rather than insight
- Avoiding commitment under the guise of caution

Presence interrupts this loop.

By grounding attention in the present moment, the mind regains proportion—and thinking becomes purposeful again.

The Body as a Source of Information

Decision-making is not purely cognitive.

The body continuously registers signals:

- Tension
- Constriction
- Ease
- Expansion
- Fatigue
- Readiness

Without presence, these signals are either ignored or mistaken for emotion.

With presence, they become **data**.

This is not intuition as impulse.
It is somatic awareness as feedback.

The body does not predict the future.
It informs you about **alignment in the present**.

Holding Multiple Truths at Once

Uncertainty often contains contradictions.

Two options may both be valid.
Two values may both matter.
Two risks may both be real.

Presence increases the capacity to hold complexity without collapse.

Instead of forcing false certainty, you can:

- Acknowledge trade-offs
- Accept partial loss
- Choose without needing to justify everything
- Move forward without closing prematurely

This capacity is emotional maturity—not indecision.

Deciding Without Emotional Avoidance

Some decisions are delayed not because they are unclear—but because they are emotionally uncomfortable.

They involve:

- Disappointment
- Conflict
- Letting go
- Being misunderstood
- Risking regret

Presence reveals when avoidance is disguised as deliberation.

By staying with the discomfort instead of escaping it, the decision clarifies on its own.

You stop asking, "What will make this easier?"
And begin asking, "What is most honest right now?"

The Timing of Decisions

Presence sharpens sensitivity to timing.

Some decisions need to be made quickly.
Others need space.

Without presence:

- Action is rushed
- Or endlessly postponed

With presence:

- You sense when enough information has arrived
- You feel when delay no longer serves
- You act without forcing momentum

Right timing is not calculated.
It is **felt**.

Responsibility Without Self-Blame

Decision-making under uncertainty carries risk.

Even the most present choice may lead to unintended consequences.

Presence allows responsibility without self-punishment.

You evaluate outcomes without collapsing into:

- Regret
- Shame
- Self-doubt
- Retrospective certainty

You learn, adjust, and continue.

Clarity is not proven by outcomes alone.
It is reflected in **how the decision was made**.

What This Chapter Leaves You With

Uncertainty does not require paralysis or haste.

When presence leads:

- Urgency loosens
- Thinking stabilizes
- Sensation informs
- Choice becomes available

You may not know the future.

But you can know yourself in the moment of choosing.

And that is enough to move forward with integrity, steadiness, and trust in your capacity to respond.

In the next chapter, we explore how presence reshapes resilience—not as endurance or toughness, but as the ability to remain open and responsive through change, loss, and disruption.

Chapter 13 — Resilience Without Hardening

Staying open through pressure, loss, and change

Resilience is often misunderstood.

It is commonly framed as toughness.
Endurance.
The ability to push through, hold on, and keep going no matter
the cost.

But this version of resilience carries a hidden price.

It hardens.

Over time, what began as strength becomes rigidity. What
once protected becomes isolating. The capacity to endure
slowly replaces the capacity to feel.

This chapter explores a different form of resilience—one
rooted not in armor, but in presence. A resilience that allows
you to remain open, responsive, and human, even through
adversity.

The Survival Model of Resilience

Most people learn resilience through necessity.

They endure stress.
They adapt to pressure.
They suppress vulnerability to function.

This survival-based resilience works—especially in
environments that reward performance over presence.

But survival resilience relies on contraction.

It tightens the body.
Narrows attention.
Restricts emotion.
Prioritizes control.

Over time, this creates a hardened inner posture:

- Less sensitivity
- Reduced empathy
- Emotional distance
- Chronic tension
- Difficulty resting or receiving support

What is praised as strength often masks **long-term depletion**.

Hardening Is Not Healing

Hardening protects against pain by reducing contact.

It limits feeling.
It dampens response.
It numbs vulnerability.

But healing does not occur through numbness.

When emotion is shut down, learning is delayed. When sensation is avoided, integration remains incomplete.

Hardening keeps you functioning—but not fully alive.

True resilience allows pain to be felt **without becoming defining**.

Presence as Adaptive Capacity

Presence changes the nature of resilience.

Instead of resisting experience, presence **meets** it.

Instead of bracing, presence allows movement.

Instead of tightening, presence stays available.

This creates adaptive resilience:

- The ability to feel without breaking
- The ability to bend without collapsing
- The ability to respond rather than endure

Presence does not eliminate pain.
It prevents pain from becoming identity.

Why Open Systems Recover Faster

In nature, the most resilient systems are not the hardest.

They are the most flexible.

Rigid structures fracture under pressure.
Adaptive structures redistribute force.

The human nervous system works the same way.

When presence is available:

- Emotion flows rather than stagnates
- Stress completes its cycle

- Recovery becomes possible
- Energy returns

Hardening traps stress inside the system.
Presence allows it to move through.

Loss Without Closure, Change Without Collapse

Some experiences do not resolve neatly.

Loss may not offer meaning.
Change may not feel chosen.
Disruption may not come with answers.

Resilience without hardening allows you to remain open even when there is no resolution.

You do not force closure.
You do not rush healing.
You do not turn pain into a lesson prematurely.

You allow experience to be incomplete—without becoming stuck.

This is emotional maturity.

The Courage to Stay Soft

Softness is often mistaken for weakness.

In reality, softness requires far more courage than armor.

It takes courage to:

- Feel grief without numbing
- Admit exhaustion without shame
- Stay open after disappointment
- Remain kind without self-betrayal
- Trust again without guarantees

Presence supports this courage.

It anchors softness in stability.

Resilience Is Not Isolation

Another hidden cost of hardening is disconnection.

When people harden, they withdraw—not always physically, but emotionally.

They stop sharing.
They stop asking.
They stop letting themselves be seen.

Presence-based resilience restores connection.

Not dependency.
Not collapse.

But mutuality.

The ability to be supported without losing agency.

Strength That Does Not Abandon the Self

Resilience without hardening allows strength to coexist with sensitivity.

You can be capable and tender.
Grounded and grieving.
Steady and uncertain.

You do not have to trade humanity for endurance.

Presence keeps you aligned with yourself—even when life is misaligned.

What This Chapter Leaves You With

Resilience does not require you to become less human.

It asks you to become more present.

When presence leads:

- Pain is felt without defining you
- Change is met without collapse
- Recovery becomes possible without armor

You do not need to harden to survive.

You need enough presence to stay **open**, adaptive, and intact—through whatever life brings.

In the next chapter, we explore how presence reshapes meaning itself—allowing purpose to emerge not from striving or certainty, but from alignment with what is real.

Chapter 14 — From Reactive Patterns to Responsive Living

How presence reshapes habits and identity

Most people believe their lives are shaped by deliberate choices.
In reality, much of life is governed by patterns formed long before choice enters awareness.

We react before we decide.
We defend before we understand.
We repeat behaviors not because they are wise, but because they are familiar.

This chapter explores how presence quietly dissolves these patterns—not by force, discipline, or self-improvement—but by altering the inner conditions from which behavior arises.

The Hidden Architecture of Reaction

Reactive patterns are not flaws of character.
They are adaptations.

At some point, reacting quickly kept you safe, accepted, or in control.
Speed became competence.
Defense became strength.
Avoidance became relief.

Over time, these responses solidified into habits.
Habits hardened into identity.

You stopped saying "I react this way."
You started saying "This is just who I am."

Presence challenges this assumption at its root.

Why Willpower Fails to Change Habits

Most attempts at change focus on behavior:

- "I should speak less sharply."
- "I need to be more patient."
- "I must stop overthinking."

But behavior is the surface expression of an inner state.

Without presence:

- Effort fights pattern.
- Discipline battles conditioning.
- Progress collapses under stress.

This is why people can change briefly—then revert the moment pressure returns.

Presence works differently.
It does not fight the habit.
It changes the *ground* from which the habit grows.

The Moment Pattern Loses Its Grip

A reactive pattern requires one thing to function:
unconsciousness.

The moment you fully see a pattern as it activates—
not intellectually, but experientially—
it loses authority.

This is not because you stopped it.
It is because you were no longer inside it.

Presence creates a subtle but decisive shift:

- From *being* the reaction
- To *witnessing* the reaction

That witnessing is not passive.
It is transformative.

From Compulsion to Choice

In reactive living:

- Stimulus → reaction → justification

In responsive living:

- Stimulus → awareness → choice

The space between stimulus and response is not created by effort.
It emerges naturally when presence is stable.

In that space:

- Old habits feel optional.
- Emotional impulses soften.
- Identity loosens.

You begin to notice something quietly liberating:

You are not required to become who you have always been.

How Presence Rewrites Identity

Identity is often mistaken for personality.
In truth, it is accumulated memory reinforced by repetition.

"I am impatient."
"I am conflict-avoidant."
"I am intense."
"I am sensitive."

Presence does not argue with these identities.
It outgrows them.

As presence deepens:

- Reactions feel less personal.
- Roles feel less fixed.
- Self-descriptions feel provisional.

You are no longer defined by what arises—
only by how consciously you meet it.

Identity shifts from who you think you are
to how present you are being.

Living Without the Old Inner Narrator

Reactive patterns are fueled by an internal narrator:

- Explaining
- Defending
- Interpreting
- Judging

Presence quiets this voice—not by silencing it, but by removing its monopoly.

When the narrator relaxes:

- Experience becomes simpler.
- Emotions pass more quickly.
- Life feels less like a performance.

You still think.
You still decide.
But thought is no longer driving from the front seat.

Responsive Living Is Not Slower—It Is Clearer

A common fear is that presence will make life passive or indecisive.

The opposite is true.

Reactive speed is urgency.
Responsive speed is clarity.

When action arises from presence:

- It is precise.
- It is proportionate.
- It is grounded.

You respond faster because you are not entangled.

The Quiet Revolution of Being Here

The shift from reactive to responsive living does not announce
itself.
There is no dramatic before-and-after moment.

Instead, you notice small but meaningful changes:

- Fewer regrets.
- Less emotional residue.
- Greater consistency under pressure.

Life begins to feel less like something happening *to* you
and more like something moving *through* you.

What Presence Ultimately Changes

Presence does not perfect behavior.
It humanizes it.

You still feel emotion.
You still encounter difficulty.
You still make mistakes.

But the struggle to control yourself dissolves.

You are no longer managing a reactive self.
You are inhabiting a responsive one.

And in that shift, habits soften, identity loosens, and life
regains its natural intelligence.

Presence is not a new way of doing.
It is a new way of being—from which doing finally makes
sense.

Chapter 15 — Presence as a Way of Being

Not something you do—but how you live

Presence is often misunderstood because we approach it the way we approach skills: something to practice, something to remember, something to apply when needed. We ask, *How do I be present right now?* as if presence were an action we switch on, a technique we perform, or a state we visit briefly before returning to normal life.

But presence is not an intervention.
It is an orientation.

It is not something you *do* in moments of stress.
It is how you *meet* moments—stressful or otherwise.

This chapter is not about adding another practice to your life. It is about recognizing what changes when presence becomes the ground from which life is lived.

From Effort to Embodiment

When presence is treated as a task, it quickly becomes exhausting. You notice yourself reminding, correcting, trying to "stay present," and judging yourself when you fail. Presence then becomes another standard to meet—another place where effort replaces ease.

True presence does not require monitoring.
It does not require self-correction.
It does not require vigilance.

It arises when effort relaxes.

This does not mean passivity. It means *alignment*. When attention is no longer divided between experience and commentary, between feeling and resistance, something settles. You are no longer managing yourself. You are inhabiting yourself.

Presence is what remains when unnecessary tension dissolves.

The Difference Between Moments of Presence and a Present Life

Many people have tasted presence—during meditation, in nature, in moments of deep connection, or after loss strips life down to what matters. These moments are meaningful, but they are temporary. Life resumes. Urgency returns. Old patterns reassert themselves.

A present life is different.

In a present life:

- You notice reactions earlier, before they harden into behavior.
- You recover faster when you do get pulled off-center.
- You no longer experience calm and chaos as opposites.
- You stop outsourcing stability to circumstances.

Presence does not eliminate difficulty.
It changes your relationship to difficulty.

The difference is subtle but profound: instead of asking, *How do I get back to presence?* you begin to notice when you *leave* it—and why.

Presence and Identity

One of the reasons presence feels elusive is that it quietly threatens identity. Much of who we think we are is maintained through inner narration: opinions, justifications, defenses, rehearsals. Presence reduces this noise—not by force, but by irrelevance.

When presence deepens, identity loosens.

You still have roles, values, and commitments—but they are no longer protected by tension. You do not need to constantly reinforce who you are. You simply respond from where you are.

This can feel unfamiliar at first. Without internal commentary, there is less drama, less self-definition, less friction. Some mistake this for emptiness or detachment. In reality, it is *contact*—direct, unfiltered contact with life as it is.

Presence does not erase personality.
It removes the strain from carrying it.

Living Without Constant Self-Interruption

Most people interrupt themselves constantly. They second-guess while speaking, anticipate while listening, and evaluate while feeling. Life is experienced through a filter of internal management.

Presence removes the need for interruption.

When you are present:

- Listening becomes full because you are not preparing a response.
- Action becomes cleaner because it is not diluted by doubt.
- Emotions move without stagnating into mood or identity.
- Silence feels supportive rather than awkward.

This does not make you slower.
It makes you *clearer*.

Life begins to feel less like a series of tasks and more like a continuous conversation—one you are actually available for.

Presence in Ordinary Life

Presence does not announce itself. It does not feel elevated or special. In fact, it often feels unremarkable—until you notice what is missing.

Less inner pressure.
Less reactivity.
Less need to control outcomes.

Ordinary moments carry more weight. A conversation, a decision, a pause between activities—these become complete in themselves, rather than stepping stones to something else.

You stop living *toward* life and start living *within* it.

This is why presence cannot be maintained through willpower. Willpower belongs to the mind that wants to get somewhere. Presence belongs to the body that is already here.

When Presence Becomes Trust

Over time, presence grows into trust—not trust that things will go your way, but trust in your capacity to meet what arises.

You trust that:

- You will notice when something matters.
- You will respond rather than react.
- You do not need to anticipate every possibility.
- You can pause without losing momentum.

This trust is not intellectual. It is lived. It is the quiet confidence that comes from not abandoning yourself under pressure.

Presence is not certainty.
It is steadiness in uncertainty.

A Way of Being, Not a Destination

There is no final state of presence. There is no permanent arrival. Life will continue to challenge attention, trigger emotion, and test clarity. The difference is not that you transcend these experiences—but that you remain *with* them.

Presence does not mean you are always centered.
It means you are always *available* to re-center.

And eventually, even that effort fades. Presence becomes less something you return to and more something you notice when it is obscured.

At that point, presence is no longer a tool.
It is a way of being human.

You do not practice it to improve life.
You live it because it is how life feels most honest, most grounded, most whole.

And from that place, emotional intelligence is no longer something you apply.
It is simply how you move through the world.

Closing Reflection

The Quiet Power of Being Fully Here

Presence is not something you add to your life.
It is what remains when you stop leaving it.

Throughout this book, we have not chased better techni

ques, stronger discipline, or more refined control. We have returned—again and again—to something simpler and far more demanding: the willingness to be here, as life is, without fleeing into reaction.

Being fully here does not mean being calm all the time.
It does not mean having answers.
It does not mean feeling certain or resolved.

It means staying.

Staying with discomfort long enough for it to soften.
Staying with uncertainty without rushing to false clarity.
Staying with emotion without being overtaken by it.
Staying with another human being without needing to fix,
defend, or perform.

This is the quiet power beneath emotional intelligence.
Not emotional management—but emotional contact.
Not control—but capacity.
Not effort—but availability.

When presence becomes the ground you stand on, something
subtle yet profound shifts. You begin to respond rather than
react—not because you are trying harder, but because there is
more space inside you. Space between stimulus and
response. Space between feeling and action. Space between
who you are and what you experience.

From that space, intelligence emerges naturally.

Decisions become clearer—not because the world simplifies,
but because your inner field is less crowded. Conversations
deepen—not because you say more, but because you are no
longer elsewhere while listening. Leadership steadies—not
because pressure disappears, but because you no longer
collapse under it.

Presence does not make life easier.
It makes you more *able*.

And perhaps most importantly, presence restores something we did not realize we had lost: a sense of being at home in ourselves. When you are fully here, even difficult moments carry dignity. Even pain is held rather than resisted. Even silence feels inhabited rather than empty.

This is not a practice you complete.
It is a way of being you return to—again and again.

You will forget.
You will rush.
You will react.

And then—if this book has done its work—you will remember. You will pause. You will come back.

That return is the work.
That return is the power.

To be fully here is not a grand achievement.
It is a quiet commitment—to meet life as it unfolds, with awareness intact and heart available.

Nothing more is required.
Nothing less will do.

Presence

The Foundation of

EMOTIONAL INTELLIGENCE

Appendices

Appendix A — What Presence Is (and Is Not)

Presence is often spoken about, rarely defined, and frequently misunderstood. This appendix clarifies what presence actually is—so it can be recognized, cultivated, and trusted—and what it is not—so it is not diluted into technique, performance, or personality trait.

This distinction matters. When presence is misunderstood, people strive for the wrong thing. When it is seen clearly, it becomes accessible in ordinary life.

What Presence Is

Presence is a state of inner availability.
It is the condition in which attention, awareness, and emotional regulation are aligned with what is actually happening—internally and externally—without distortion, urgency, or avoidance.

Presence is not effortful. It is not something you "do" so much as something that emerges when resistance drops.

At its core, presence includes:

1. Awareness without interference
You notice sensations, thoughts, and emotions as they arise—without immediately trying to fix, suppress, explain, or justify them. Awareness precedes reaction.

2. Regulation without suppression

The nervous system is sufficiently settled to allow choice. Emotions are felt without overwhelming the system or being pushed away.

3. Attention that is grounded, not narrow

You are engaged with what matters now, without tunnel vision. Focus is held within a wider field of awareness.

4. Contact with reality as it is

Presence meets the moment before preferences, stories, or defenses reshape it. This does not mean passivity—it means accuracy.

5. Choice between stimulus and response

Presence creates space. In that space, intelligence operates.

Presence is subtle. It is often quiet. It may not feel dramatic— but its effects are profound.

What Presence Is Not

Many practices and traits resemble presence on the surface but are fundamentally different in function.

Presence is not calmness.
Calm may arise from presence, but presence can exist amid intensity, grief, urgency, or conflict. Presence does not require a particular emotional tone.

Presence is not relaxation.
Relaxation is a bodily state. Presence is a relational state—

with yourself, others, and the moment. You can be present while energized or under pressure.

Presence is not mindfulness as a task.
Techniques can support presence, but presence itself is not an exercise to perform. When it becomes mechanical, it loses its essence.

Presence is not emotional suppression.
Being composed by shutting down feeling is not presence—it is disconnection. Presence includes emotional contact, not emotional absence.

Presence is not positivity.
Presence does not replace difficult emotions with pleasant ones. It allows the full spectrum of experience without distortion.

Presence is not personality.
Some people appear naturally grounded or attentive. Presence is not a trait you have or lack—it is a capacity available to every human nervous system.

Presence is not control.
Trying to control thoughts, emotions, or outcomes pulls you out of presence. Presence allows influence precisely because it releases control.

Why This Distinction Matters

When presence is confused with calmness, people believe they have "lost" presence during stress.

When it is confused with technique, people strive harder—and become more fragmented.
When it is confused with positivity, people avoid necessary emotions.

True presence is compatible with complexity. It does not disappear under pressure—it becomes more essential.

A Simple Test

You can recognize presence through three signals:

- You are aware of what you feel without being consumed by it
- Your attention is steady but flexible
- Your responses feel proportionate, not reactive

If these are present, presence is present—regardless of what you are feeling.

The Quiet Paradox

Presence cannot be forced, yet it can be invited.
It cannot be performed, yet it can be practiced.
It cannot be held permanently, yet it returns the moment resistance softens.

Presence is not an ideal state to maintain.
It is a home you can return to—again and again—whenever you notice that you have left.

This appendix is not an endpoint.
It is a compass—so that as you continue living, leading, deciding, and relating, you know what you are actually cultivating beneath every practice in this book.

Appendix B — The Presence–Emotional Intelligence Map

ost emotional intelligence frameworks describe *skills*: self-awareness, self-regulation, empathy, social effectiveness, decision-making.
What they often leave unspoken is the *state* from which these skills arise.

This appendix makes that foundation explicit.

Presence is not one component of emotional intelligence.
It is the ground from which every component functions—or collapses.

What follows is a conceptual map showing how presence enables each core dimension of emotional intelligence, and what happens when presence is absent.

---

1. Presence as the Foundational State

Presence is the capacity to remain here—internally and externally—without being hijacked by urgency, fear, identity defense, or emotional noise.

It is not calm.
It is not control.
It is not effort.

It is stability of awareness under conditions of pressure.

When presence is intact:

- Perception is wider
- Emotions are felt without overwhelm
- Choice precedes reaction

When presence collapses:

- Attention narrows
- Emotion becomes directive
- Habit replaces choice

Every emotional intelligence competency sits *downstream* of this state.

---

2. Self-Awareness: Seeing Clearly Without Judgment

With Presence

- Sensations, emotions, and thoughts are noticed in real time
- Awareness precedes interpretation
- Identity is not fused with experience ("I feel anger" vs. "I am angry")

Without Presence

- Awareness becomes retrospective ("I realize later what happened")
- Emotion is noticed only after action
- Self-awareness collapses into self-criticism or justification

Map Insight

Presence creates the *pause* that allows awareness to arise without distortion.

---

3. Self-Regulation: Stability Without Suppression

With Presence

- Emotions move through without needing to be fixed
- Regulation happens through allowance and choice
- Energy is conserved rather than drained

Without Presence

- Regulation becomes control or repression
- Emotional energy is spent resisting what is already present
- Suppressed emotion resurfaces under pressure

Map Insight

Presence regulates by *holding*, not by pushing away.

---

4. Empathy: Openness Without Self-Loss

With Presence

- Another's emotional field is perceived without absorption
- Compassion arises without over-identification
- Boundaries remain intact

Without Presence

- Empathy turns into emotional contagion
- Other people's emotions hijack internal state
- Burnout, withdrawal, or detachment follow

Map Insight

Presence allows empathy *with* boundaries, not empathy *as* boundary loss.

---

5. Social Awareness & Relationship Intelligence

With Presence

- Subtle cues are perceived: tone, timing, silence
- Listening is receptive, not strategic
- Responses are attuned rather than rehearsed

Without Presence

- Attention is split between self-protection and performance
- Listening becomes transactional
- Misattunement increases conflict

Map Insight

Presence restores relational intelligence by freeing attention from self-monitoring.

---

6. Decision-Making: Choice Under Uncertainty

With Presence

- Multiple signals are held simultaneously

- Emotion informs but does not dictate
- Clarity emerges even without certainty

Without Presence

- Decisions are rushed to relieve discomfort
- Fear, urgency, or ego dominate
- Regret often follows action

Map Insight

Presence allows decisions to arise from integration, not pressure.

---

7. Motivation & Drive: Action Without Strain

With Presence

- Action is aligned rather than compulsive
- Energy is sustainable
- Purpose guides effort

Without Presence

- Drive becomes proving, defending, or escaping
- Burnout replaces fulfillment
- Achievement feels hollow

Map Insight

Presence shifts motivation from *tension-based* to *meaning-based* action.

---

8. The Complete Map (Summary View)

Presence enables:

- Awareness → before reaction
- Regulation → without suppression
- Empathy → without exhaustion
- Connection → without performance
- Decisions → without urgency
- Action → without self-betrayal

Emotional intelligence does not fail because people lack knowledge.
It fails because presence collapses under pressure.

A Final Reframe

Emotional intelligence is not something you *apply* when things are calm.
It is something that *emerges* when presence is stable.

Strengthening presence strengthens every dimension of emotional intelligence simultaneously—without adding techniques, scripts, or strategies.

This is the map beneath the methods.

Appendix C — Signals of Presence and Absence

How inner state quietly reveals itself in behavior, attention, and impact

Presence is not an idea people hold—it is a condition others *feel*.
Absence is not a moral failure—it is a nervous system under load.

This appendix offers a practical way to **recognize presence and absence as signals**, not labels. These signals appear in posture, speech, timing, listening, and emotional tone. They shape trust, clarity, and influence—often before a single word lands.

Use this appendix as a mirror, not a measurement.
The purpose is awareness, not self-judgment.

Why Signals Matter

Most people try to *perform* emotional intelligence.
Presence makes performance unnecessary.

Signals of presence and absence:

- Appear before content
- Are sensed faster than logic
- Influence outcomes **more than intention**

People rarely remember *what* was said—but they remember **how it felt to be with you**.

Core Distinction

Presence

Attention is here. The nervous system is regulated. Awareness is wide.

Absence

Attention is fragmented. The nervous system is defensive. Awareness is narrow.

Everything else flows from this difference.

Signals of Presence

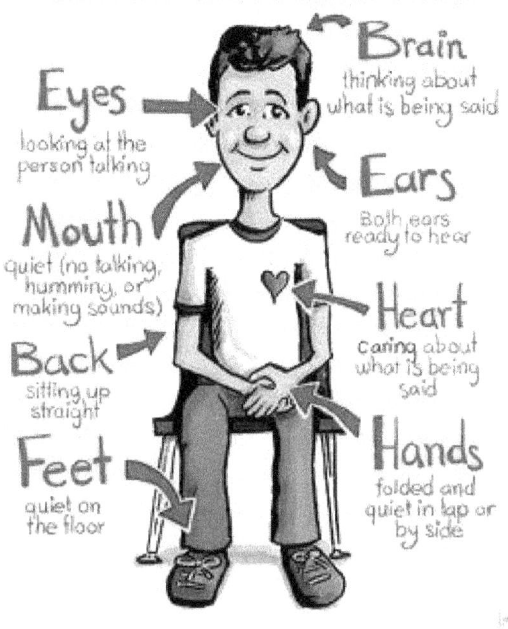

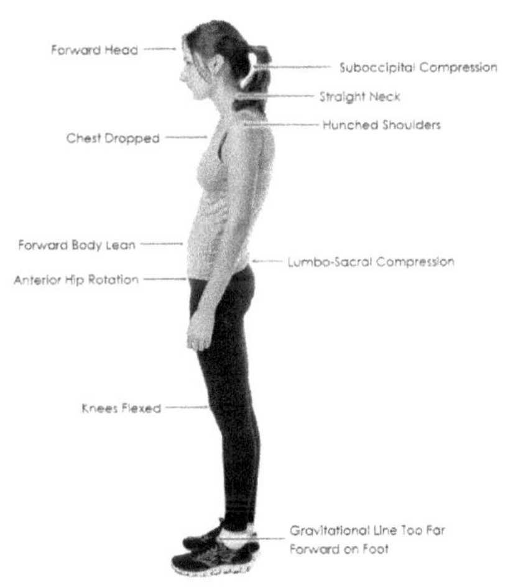

BAD POSTURE

GOOD POSTURE

These signals arise *naturally* when awareness is steady and the nervous system is settled.

1. Attention Signals

- Listens without rehearsing responses
- Stays with one conversation at a time
- Notices subtle emotional shifts
- Responds to what *is actually said*, not what was expected

Felt experience: "I am being met."

---

2. Emotional Signals

- Emotions move without spilling or suppression
- Calm exists without detachment
- Empathy without urgency to fix
- Stability without rigidity

Felt experience: "I can relax here."

---

3. Behavioral Signals

- Speech is measured, not rushed
- Pauses appear naturally
- Decisions feel grounded, not forced
- Boundaries are clear and kind

Felt experience: "This person is steady."

---

4. Relational Signals

- Others speak more honestly
- Conflict de-escalates without strategy
- Trust builds without explanation
- Silence feels safe, not awkward

Felt experience: "I don't need to defend myself."

---

5. Leadership Signals (when applicable)

- Authority without dominance
- Confidence without performance
- Direction without pressure
- Accountability without fear

Felt experience: "I know where I stand."

Signals of Absence

These signals arise when the nervous system is overloaded or identity is threatened.

1. Attention Signals

- Interrupts or finishes others' sentences
- Multitasks while "listening"
- Misses emotional subtext
- Reacts to assumptions rather than reality

Felt experience: "I'm not really being heard."

2. Emotional Signals

- Irritability or emotional flatness
- Defensiveness disguised as certainty
- Empathy fatigue
- Mood shifts without awareness

Felt experience: "I need to be careful here."

3. Behavioral Signals

- Speech feels rushed or excessive
- Over-explaining or withdrawing
- Compulsive problem-solving
- Avoidance of discomfort

Felt experience: "Something feels off."

4. Relational Signals

- Conversations feel transactional
- Misunderstandings escalate quickly
- People self-censor
- Trust erodes quietly

Felt experience: "I don't feel safe being honest."

5. Leadership Signals (when applicable)

- Control replaces clarity
- Urgency replaces direction
- Compliance replaces commitment
- Fear replaces accountability

Felt experience: "I'm managing reactions, not meaning."

A Crucial Clarification

Presence and absence are **states**, not traits.

- Everyone moves between them
- Stress increases absence
- Safety restores presence
- Awareness is the turning point

The moment absence is *noticed*, presence begins to return.

The Most Reliable Indicator

The clearest signal is not behavior—it is **impact**.

Ask quietly:

- Do people open or close around me?
- Does clarity increase or decrease?
- Does tension soften—or spread?

These outcomes reveal presence more accurately than self-assessment.

Using This Appendix Practically

Try this simple reflection at the end of a day or conversation:

1. Where did presence feel *natural* today?
2. Where did absence quietly appear?
3. What conditions supported or disrupted regulation?
4. What would restore steadiness next time?

No fixing required.
Only noticing.

Closing Note

Presence does not announce itself.
It signals quietly—through steadiness, spaciousness, and safety.

Absence does not mean failure.
It means the system needs care, not correction.

This appendix is not a checklist to perfect—but a compass to return.

Appendix D — Micro-Practices for Presence

mall moments. Immediate shifts. Sustainable change.

Presence is not built through long retreats or perfect conditions.
It is shaped through **brief, repeatable moments** woven into ordinary life.

Micro-practices work because they meet reality as it is—busy, imperfect, emotionally charged. Each practice in this appendix requires **30 seconds to 3 minutes** and can be used anywhere: at work, in conversation, in traffic, or at home.

These are not techniques to master.
They are **entry points back into being here.**

---

1. The Arrival Pause (15–30 seconds)

When to use: Before entering a meeting, conversation, or task.

How:

- Stop just before crossing the threshold.
- Feel both feet on the ground.
- Take one slow breath, noticing the exhale.
- Silently say: *"I arrive."*

Why it works:
It interrupts momentum and signals the nervous system that the next moment deserves full presence.

---

2. One-Breath Reset (10–15 seconds)

When to use: During stress, irritation, or urgency.

How:

- Inhale gently through the nose.
- Exhale slightly longer than the inhale.
- Let the shoulders drop on the out-breath.

Why it works:
A single elongated exhale activates regulation faster than cognitive effort.

---

3. The Sensory Anchor (30–60 seconds)

When to use: When attention scatters or emotions escalate.

How:

- Name silently:
  - 1 thing you can see
  - 1 thing you can hear
  - 1 sensation in the body

Why it works:
Presence returns through the senses—not analysis.

---

4. The Inner Posture Check (20 seconds)

When to use: In conversations or decision-making.

How:
Ask quietly:

- "Am I open or defended?"
- "Am I here to understand or to control?"

No fixing. Just noticing.

Why it works:
Awareness of posture dissolves unconscious reactivity.

---

5. The Name-the-State Practice (15 seconds)

When to use: When emotions arise suddenly.

How:
Silently name the state:

- Tightness
- Urgency
- Irritation
- Fatigue

Why it works:
Naming creates space between awareness and reaction.

---

6. Micro-Grounding Through the Body (30 seconds)

When to use: When thoughts loop or speed up.

How:

- Press your feet gently into the floor.

- Feel contact points—chair, hands, back.
- Let attention rest there.

Why it works:
The body is always in the present moment.

---

7. The Three-Word Check-In (10 seconds)

When to use: Throughout the day.

How:
Ask:
"What three words describe my state right now?"

Why it works:
It restores self-awareness without overthinking.

---

8. Listening Without Planning (1–3 minutes)

When to use: In conversations.

How:

- Listen without preparing your response.
- Notice tone, pace, and emotion—not just words.

Why it works:
Presence deepens connection more than clever replies.

---

9. The Pause Before Response (5 seconds)

When to use: When triggered or challenged.

How:

- Pause before speaking.
- Feel one breath.
- Respond from clarity, not impulse.

Why it works:
This is where emotional intelligence becomes visible.

---

10. Closing the Moment (20 seconds)

When to use: After meetings, interactions, or tasks.

How:

- Take one breath.
- Acknowledge: "This moment is complete."

Why it works:
Unfinished emotional loops drain presence across the day.

How to Use These Practices

- Choose **one or two** to use consistently.
- Practice **in calm moments**, not only under stress.
- Let frequency matter more than duration.
- Return gently—without judgment—whenever you forget.

Presence grows not by effort, but by returning again and again.

Final Note

Micro-practices are not tools to fix yourself.
They are reminders of what is already available.

Presence is not created.
It is remembered—one small moment at a time.

Appendix E — Presence Under Pressure

Staying grounded when it matters most

Pressure is not the enemy of presence.
Unexamined pressure is.

This appendix brings together the core truth that runs through this book: **presence is not proven in calm moments—it is revealed under strain**. When time compresses, emotions intensify, and stakes rise, presence does not disappear. What disappears is our access to it.

This appendix clarifies why presence collapses under pressure—and how it can be restored, even in the most demanding situations.

---

1. Why Pressure Narrows Presence

Under pressure, the nervous system prioritizes survival over awareness.

This creates three predictable shifts:

- **Attention contracts** toward threat, urgency, or outcomes
- **Time perception compresses**, creating a sense of "now or never"
- **Identity tightens**, as the ego tries to protect competence, image, or control

In this state:

- Listening becomes selective
- Thinking becomes binary
- Emotions become reactive rather than informative

Presence is not lost—it is overridden.

---

2. The Illusion of "Handling Pressure Well"

Many people pride themselves on functioning under pressure. But performance is not the same as presence.

Common pressure adaptations include:

- Emotional suppression masked as professionalism
- Speed mistaken for decisiveness
- Control mistaken for leadership
- Detachment mistaken for composure

These strategies may deliver short-term results—but they extract a long-term cost:

- Reduced emotional sensitivity
- Narrower perspective
- Delayed burnout
- Relational erosion

True presence does not harden under pressure.
It **stays permeable without collapsing**.

---

3. Pressure Is a State—Not a Situation

Pressure is not caused by events.
It is created by **internal interpretation + nervous system response**.

Two people can face the same situation:

- One becomes reactive and rigid
- The other becomes focused and grounded

The difference is not resilience alone.
It is **regulation plus awareness**.

Presence under pressure begins when pressure is recognized as an internal state—something that can be worked with, not fought.

---

4. The Three Signals of Presence Collapse

You can detect loss of presence under pressure through three early signals:

1. **Urgency replaces clarity**
   Everything feels immediate, but nothing feels grounded.
2. **Defensiveness replaces curiosity**
   Questions feel threatening rather than informative.
3. **Reactivity replaces choice**
   Responses feel automatic, justified, and irreversible.

These signals are not failures.
They are **alerts**.

---

5. The Micro-Restoration of Presence

Presence does not return through effort.
It returns through **interruption**.

Under pressure, restoration happens in micro-moments:

- One conscious exhale
- One softening of the jaw or shoulders
- One deliberate pause before speaking
- One silent acknowledgment: *"I am activated."*

These are not techniques to calm down.
They are **doorways back to awareness**.

Presence returns not when pressure ends—but when
awareness re-enters.

---

6. Leading From Presence Under Pressure

In leadership contexts, presence under pressure creates a
distinct emotional field:

- Others feel steadier without knowing why
- Conversations slow without losing momentum
- Decisions become clearer, not louder
- Authority emerges without force

This is not charisma.
It is nervous system coherence.

People do not follow certainty.
They follow **regulated presence**.

7. Pressure as a Training Ground

Pressure is not an obstacle to presence.
It is its proving ground.

Every difficult conversation
Every moment of uncertainty
Every emotionally charged interaction

is an invitation to practice:

- Staying connected rather than correct
- Staying responsive rather than reactive
- Staying human rather than defended

Presence under pressure is not about mastery.
It is about **returning—again and again—to the here and now**.

Closing Reflection

Presence does not ask you to eliminate pressure.
It asks you to meet it without abandoning yourself.

When pressure rises, the question is not:
"How do I get through this?"

It is:
"Can I stay here while this moves through me?"

That is the quiet power of presence under pressure.

Appendix F — Presence in Leadership Roles

tability, clarity, and trust as lived qualities—not techniques

Leadership does not begin with strategy, authority, or communication skill.
It begins with **state**.

Before a leader speaks, decides, or directs, something quieter is already shaping the room:
their level of presence.

This appendix explores how presence functions across leadership roles—not as a performance skill, but as the underlying condition that determines credibility, influence, and emotional safety.

---

1. Leadership Is Transmitted Before It Is Expressed

People do not respond first to what a leader says.
They respond to **how regulated, grounded, and available the leader is**.

Presence is communicated through:

- Pace of speech
- Tone and breath
- Eye contact and listening
- Emotional steadiness under pressure
- Willingness to pause before reacting

When presence is strong:

- Others feel settled, even in uncertainty
- Conversations slow without losing momentum
- Authority feels natural, not enforced

When presence collapses:

- Urgency replaces clarity
- Control substitutes for leadership
- Trust erodes quietly, long before results decline

---

2. The Leader as Emotional Regulator

Whether consciously or not, leaders function as **emotional reference points**.

Teams orient themselves around:

- How the leader handles stress
- How mistakes are met
- Whether uncertainty is tolerated or punished
- Whether emotions are acknowledged or avoided

Presence allows a leader to:

- Absorb intensity without amplifying it
- Name what is happening without dramatizing it
- Stay responsive instead of reactive

This does not require emotional neutrality.
It requires **emotional capacity**—the ability to feel without being overwhelmed.

---

3. Presence Across Leadership Contexts

Presence shows up differently depending on role, but its foundation remains the same.

Executive Leadership

- Holding long-term perspective under pressure
- Making decisions without emotional narrowing
- Allowing silence before conclusions

Presence here creates **strategic calm**.

People Management

- Listening without preparing responses
- Addressing issues without defensiveness
- Remaining steady during emotional conversations

Presence here creates **psychological safety**.

Project & Operational Leadership

- Responding to problems without blame
- Maintaining clarity during disruption
- Prioritizing wisely under time pressure

Presence here creates **execution stability**.

Crisis Leadership

- Slowing the system when fear accelerates it
- Communicating simply and truthfully
- Regulating self before attempting to regulate others

Presence here creates **trust under threat**.

4. Why Presence Cannot Be Delegated

Leaders often attempt to compensate for absence with:

- More meetings
- More control
- More communication
- More policies

But presence cannot be replaced by structure.

A leader may delegate tasks, decisions, and authority—but **presence must be embodied**.

No system can override:

- A reactive nervous system
- An emotionally unavailable leader
- A chronically rushed inner state

The leader's inner condition becomes the organization's emotional climate.

5. Authority Without Presence Feels Like Pressure

When presence is absent:

- Authority becomes force
- Direction becomes demand
- Accountability becomes fear-driven

When presence is present:

- Authority feels grounded
- Direction feels clarifying
- Accountability feels fair and human

This is why some leaders command compliance but not commitment—
and others inspire trust without raising their voice.

---

6. Presence as the Foundation of Ethical Leadership

Ethical failures rarely begin with bad intentions.
They begin with **disconnection**.

Presence supports ethical leadership by:

- Slowing impulsive decisions
- Allowing discomfort without avoidance
- Creating space for reflection
- Noticing early warning signals

A present leader can feel the misalignment before it becomes misconduct.

---

7. Cultivating Presence as a Leader

Presence in leadership is not built through personality change.
It is built through **capacity expansion**.

Key practices include:

- Regular pauses before high-stakes decisions
- Conscious regulation during conflict

- Short moments of grounding between interactions
- Willingness to not know immediately

These are not time-consuming practices.
They are **state-shifting ones**.

---

8. The Quiet Impact of Present Leadership

The most powerful leaders are often described in simple terms:

- "They made me feel heard."
- "Things felt calmer when they walked in."
- "Even hard conversations felt manageable."

These are not compliments of charisma.
They are signals of presence.

Closing Reflection

Leadership is not primarily about moving people faster.
It is about **helping people move more clearly**.

Presence does not remove complexity, pressure, or responsibility.
It allows a leader to meet them **without fragmentation**.

And in doing so, it quietly reshapes:

- Decisions
- Relationships
- Culture

- Results

Not through force—but through being fully here.

Appendix G — Presence and Emotional Boundaries

Clarity without distance. Care without collapse.

Emotional boundaries are often misunderstood. They are commonly framed as defenses—ways to protect oneself from others' emotions, demands, or expectations. But true emotional boundaries are not built through withdrawal, rigidity, or emotional numbing. They emerge naturally from **presence**.

When presence is stable, boundaries do not need to be enforced. They are *felt*.

The Boundary Problem Beneath Emotional Intelligence

Many people struggle with one of two patterns:

- **Over-permeability** — absorbing others' emotions, moods, and pressures as one's own
- **Over-fortification** — emotional distancing, detachment, or guardedness mistaken for strength

Both are signs of **absence**, not balance.

When presence collapses:

- Empathy turns into emotional leakage
- Responsibility becomes over-functioning
- Compassion slides into self-erasure
- Detachment replaces discernment

Presence restores the middle ground.

What Presence Does to Boundaries

Presence creates internal definition.

When you are present:

- You know where your emotions end and another's begin
- You can feel concern without carrying burden
- You can listen deeply without absorbing distress
- You can say no without defensiveness
- You can say yes without obligation

Boundaries stop being rules you apply and become **clarity you embody**.

Boundaries Are a Function of Self-Contact

Healthy boundaries require one thing above all else: **ongoing contact with your own internal state.**

Presence provides that contact.

Without presence:

- You notice discomfort too late
- You override signals in order to please, perform, or keep peace
- You confuse urgency with responsibility
- You outsource your internal authority

With presence:

- Signals are noticed early
- Limits are felt before resentment builds
- Choices are made consciously, not reactively

Boundaries become proactive rather than corrective.

Presence Separates Empathy from Enmeshment

Empathy does not require emotional merging.

Presence allows you to:

- Be *with* someone's pain without being *inside* it
- Validate emotions without fixing them
- Care without rescuing
- Stay grounded when others are dysregulated

This distinction is subtle but transformative.

You are not responsible for regulating others.
You are responsible for **regulating yourself while remaining available**.

The Nervous System and Boundary Integrity

Emotional boundary failure often originates in the nervous system, not character.

When the nervous system is in threat:

- It seeks safety through compliance or control

- It loses the ability to distinguish self from other
- It reacts before discernment is possible

Presence stabilizes the nervous system, restoring:

- Choice
- Perspective
- Internal containment

From this state, boundaries are not defended—they are *obvious*.

Boundaries Without Presence Become Hard or Fragile

Without presence:

- Boundaries become rigid rules enforced through distance
- Or porous agreements overridden by guilt

With presence:

- Boundaries are flexible yet firm
- Context-sensitive rather than absolute
- Calmly communicated rather than justified

Presence gives boundaries **texture**, not tension.

A Quiet Truth About Emotional Strength

Strong boundaries are not loud.

They do not announce themselves.
They do not require explanation.
They do not escalate.

They are felt in:

- The calm tone of a refusal
- The ease of staying grounded in disagreement
- The absence of internal conflict after saying no
- The ability to remain open without self-betrayal

This is presence in action.

A Closing Reflection

Presence is the invisible architecture of healthy emotional boundaries.

It allows you to remain human without being overwhelmed, compassionate without being consumed, and engaged without being entangled.

When presence is steady, boundaries are not walls.

They are **clarity**.

Appendix H — Personal Presence Assessment

Presence cannot be measured like productivity or skill.
But it *can* be sensed, reflected upon, and steadily
strengthened.

This assessment is not a test.
There are no scores to chase, no ideal profile to achieve.

Its purpose is simple:
to help you notice where presence is already stable—and
where it quietly collapses.

Answer honestly. Pause between questions. Let your body
respond before your mind explains.

How to Use This Assessment

- Read each statement slowly
- Rate yourself based on **how you typically are**, not how
  you wish to be
- Use the following scale:

1 — Rarely true
2 — Occasionally true
3 — Often true
4 — Consistently true

There is no benefit in inflating your responses.
Clarity begins with truth.

Section I — Internal Awareness

These statements explore how present you are with your *inner experience*.

1. I notice tension in my body before it turns into emotion.
2. I am aware of my breathing during challenging moments.
3. I can name what I am feeling without needing to justify it.
4. I sense when I am rushing internally, even if I appear calm.
5. I recognize emotional shifts as they arise, not after they pass.

Reflection Prompt:
Where do you feel most disconnected from your internal signals?

Section II — Emotional Regulation

These statements assess how presence supports emotional stability.

6. I pause naturally before reacting when emotions are triggered.
7. I stay connected to myself even when others are upset.
8. I can feel strong emotions without needing to suppress or discharge them.

9. I recover my balance quickly after emotional
    disruptions.
10. I respond with choice rather than habit under pressure.

Reflection Prompt:
Which emotions tend to reduce your presence most quickly?

Section III — Relational Presence

These statements explore how fully you show up with others.

11. I listen without preparing my response in advance.
12. People feel heard when they speak with me.
13. I remain grounded when conversations become tense.
14. I notice when I withdraw emotionally during
    interactions.
15. I maintain connection without absorbing others'
    emotions.

Reflection Prompt:
In which relationships does your presence feel most stable—
and why?

Section IV — Presence Under Pressure

These statements focus on how stress affects your state of
being.

16. I remain centered when outcomes are uncertain.
17. I can slow down internally even when external demands
    are high.

18. I notice when fear narrows my thinking.

19. I stay connected to values rather than urgency.

20. I regain presence quickly after moments of reactivity.

Reflection Prompt:

What situations most reliably pull you out of presence?

Section V — Identity & Way of Being

These statements examine whether presence is becoming embodied.

21. I experience presence as a state, not a technique.
22. I notice when I am performing rather than being authentic.
23. I feel grounded even when nothing is being accomplished.
24. I trust stillness as much as action.
25. I relate to life from steadiness rather than effort.

Reflection Prompt:

Where are you still trying to *do* presence instead of *being* present?

Interpreting Your Responses

Do not total your score.

Instead, look for **patterns**:

- Which sections felt easy to answer?

- Which triggered discomfort or uncertainty?
- Where did you hesitate the most?

Those areas are not weaknesses.
They are **entry points for deeper presence.**

Presence Growth Indicators

As presence strengthens, you may notice:

- Shorter recovery time after emotional triggers
- Increased clarity without force
- Greater emotional resilience with less effort
- More spacious listening
- Reduced need to control outcomes

These changes are subtle—but unmistakable.

Closing Reflection

Presence does not demand improvement.
It invites honesty.

Return to this assessment periodically—not to track progress,
but to **re-establish relationship with yourself.**

Presence grows not through striving,
but through the courage to stay.

Appendix I — A 14-Day Presence Integration Path

A lived transition from understanding presence to inhabiting it

This 14-day path is not a program to complete, master, or optimize.
It is a **gentle recalibration**—a way of letting presence become familiar in your body, your attention, and your relationships.

Each day introduces **one orientation**, **one brief practice**, and **one reflective inquiry**.
The practices are intentionally simple. Their power lies in **consistency**, not intensity.

Do not rush. Do not accumulate.
If you miss a day, resume without judgment.

Presence deepens through *returning*.

How to Use This Path

- Spend 5–10 minutes per day
- Choose a consistent time
- Read the day's orientation slowly
- Practice once—or revisit it across the day
- Reflect briefly in writing or silent awareness

WEEK ONE — STABILIZING AWARENESS

Learning to arrive without force

Day 1 — Arriving Where You Already Are

Orientation: Presence begins by stopping the habit of arriving elsewhere.
Practice: Pause. Feel your feet. Feel your breath without changing it. Name silently: *Here.*
Inquiry: What shifts when I stop trying to arrive somewhere better?

Day 2 — Noticing Without Fixing

Orientation: Awareness does not require improvement.
Practice: Observe one sensation or emotion for 60 seconds without labeling it good or bad.
Inquiry: What do I habitually try to correct the moment I notice it?

Day 3 — Separating Awareness From Thought

Orientation: Thoughts are content; presence is context.
Practice: When a thought arises, note gently: *thinking.* Return to sensation.
Inquiry: Who is aware of my thoughts?

Day 4 — Sensing the Body as Anchor

Orientation: Presence stabilizes in the body before it clarifies the mind.
Practice: Scan from head to feet slowly. Do not interpret—only sense.
Inquiry: Where does my attention resist settling?

Day 5 — Allowing What Is

Orientation: Resistance amplifies noise; allowance restores space.
Practice: Let one uncomfortable sensation exist without managing it.
Inquiry: What happens when I allow instead of control?

Day 6 — Staying With the Breath

Orientation: Breath is not a tool—it is a meeting point.
Practice: Follow three full breath cycles. No modification.
Inquiry: What does the breath teach me about effort?

Day 7 — Presence Without Performance

Orientation: You are not practicing presence for results.
Practice: Sit quietly for two minutes with no objective.
Inquiry: Who am I when nothing needs to be achieved?

WEEK TWO — INTEGRATING PRESENCE INTO LIFE

Letting presence move beyond practice

Day 8 — Presence in Movement

Orientation: Presence is portable.
Practice: Walk slowly for one minute, sensing contact and motion.
Inquiry: How often do I leave my body while moving?

Day 9 — Presence in Listening

Orientation: Listening reveals presence more than silence does.
Practice: Listen to one person without preparing your response.
Inquiry: What changes when I listen to understand, not reply?

Day 10 — Presence Under Mild Stress

Orientation: Presence is most needed where it is least comfortable.
Practice: During a small stressor, feel the body before reacting.
Inquiry: Where do I lose presence first under pressure?

Day 11 — Choosing Response Over Reaction

Orientation: The pause creates freedom.

Practice: Insert one conscious breath before responding today.

Inquiry: What becomes possible in that pause?

Day 12 — Boundaries as Presence

Orientation: Presence includes knowing when not to engage.

Practice: Notice one moment where you say "no" internally.

Inquiry: How does presence protect my energy?

Day 13 — Compassion Without Collapse

Orientation: Presence allows care without absorption.

Practice: Feel empathy while remaining aware of your own body.

Inquiry: What helps me stay open without losing myself?

Day 14 — Presence as a Way of Being

Orientation: Presence is not something you do—it is how you meet life.

Practice: Reflect quietly on the last 14 days. No evaluation.

Inquiry: What has subtly changed in how I relate to myself?

After the 14 Days

You have not completed presence.
You have **established familiarity**.

Return to any day.
Repeat the path slowly.
Let presence deepen by living it—especially when it feels inconvenient.

The real integration happens **off the page**.

Presence grows where you stop trying to grow it.

Appendix J — Presence Language Guide

How words either stabilize awareness—or quietly pull us back
into reactivity

Language is not neutral.
Every word carries *state*.

Some words tighten the nervous system, accelerate thinking,
and reinforce urgency.
Others slow perception, widen awareness, and restore choice.

This guide is not about speaking *nicely* or *politely*.
It is about speaking from **presence**—language that reflects
grounded awareness rather than emotional reflex.

Use this appendix as a reference, a practice, and a mirror.

I. The Principle: Language Reveals Inner State

Before language influences others, it reveals *you*.

- Reactive language contracts time ("now," "always,"
  "never")
- Present language expands space ("here," "this," "right
  now")
- Ego-defensive language protects identity
- Presence-based language protects clarity

Presence language does **not**:

- Control outcomes
- Avoid truth

- Soften reality

It **does**:

- Slow escalation
- Preserve dignity
- Keep perception wide

II. Reactive Language vs Presence Language

Reactive Pattern	Presence-Based Alternative
"You always…"	"What I'm noticing right now is…"
"This is unacceptable."	"This doesn't work for me."
"We need to fix this immediately."	"Let's pause and see what's actually needed."
"That's wrong."	"I see this differently."
"You're not listening."	"I don't feel heard yet."
"Why did you do that?"	"Can you help me understand what led to this?"

Shift observed:

From accusation → observation
From certainty → curiosity
From pressure → clarity

III. Language That Signals Presence

These phrases subtly regulate emotional tone—both yours and others'.

Grounding phrases

- "Let me pause for a moment."
- "I want to respond thoughtfully."
- "Here's what feels clear to me."

Orientation phrases

- "Right now, what matters most is…"
- "From where I'm standing…"
- "At this moment…"

Boundary phrases

- "I'm not available for this tone."
- "I can continue if we slow this down."
- "I need a moment before responding."

Presence language sets boundaries **without threat**.

IV. Words That Collapse Presence (and Why)

Certain words reliably narrow perception:

- **Always / Never** → Freeze perspective into absolutes
- **Should / Must** → Trigger compliance or resistance
- **Why (when emotional)** → Activates defense, not insight
- Obviously / Everyone knows → Shames awareness

Replace absolutes with *specifics*.
Replace demands with *clarity*.

V. Listening Language (Often Overlooked)

Presence is heard as much in listening as in speaking.

Presence-signaling responses

- "I'm with you."
- "Go on."
- "Take your time."
- "That makes sense given what you're dealing with."

Non-presence responses (even if well-intended)

- "At least..."
- "You should just..."
- "That reminds me of..."

Listening language keeps the *center of gravity* with the speaker.

VI. Presence in Difficult Conversations

When emotions rise, presence language becomes simpler—not more complex.

Use

- Short sentences
- Neutral tone
- Fewer explanations

Examples:

- "I'm feeling overwhelmed."
- "Let's pause."
- "This matters to me."
- "I'm not ready to decide yet."

Complex justifications invite debate.
Simple presence invites regulation.

VII. Inner Language Matters First

The most influential language is the one no one hears.

Reactive inner language

- "I'm failing."
- "This shouldn't be happening."
- "I have to fix this."

Presence-based inner language

- "This is uncomfortable—and manageable."
- "I can stay with this."
- "Clarity will come."

Change inner language → outer language follows naturally.

VIII. Daily Practice: One-Line Presence

Choose **one** phrase to practice daily:

- "Let me pause before responding."
- "I'm noticing my reaction."

- "What's actually needed right now?"

Repeat it internally *before* speaking.

Language practiced under calm conditions becomes accessible under pressure.

Closing Reflection

Presence language does not perform.
It does not persuade.
It does not impress.

It stabilizes.

And from stability, truth can be spoken without harm,
boundaries can be set without force,
and connection can exist without self-loss.

Words don't just express presence.
They **transmit** it.